Digital Electronics: A Practical Approach

With EASY-PC and PULSAR

Richard Monk

Newnes

An imprint of Butterworth-Heinemann

Newnes
An imprint of Butterworth-Heinemann
Linacre House, Jordan Hill, Oxford OX2 8DP
225 Wildwood Avenue, Woburn, MA 01801–2041
A division of Reed Educational and Professional Publishing Ltd

 A member of the Reed Elsevier plc group

OXFORD   BOSTON   JOHANNESBURG
MELBOURNE   NEW DELHI   SINGAPORE

First published 1998

British Library Cataloguing in Publication Data
Monk, Richard
  Digital electronics: a practical approach with EASY PC and PULSAR
  1   Digital electronics – Data processing
  I   Title
  621.3'81'0285

ISBN 0 7506 3099 X

Typeset by Laser Words, Madras, India
Printed and bound in Great Britain

Contents

Preface

Aim

This book has two main threads. Firstly it provides a practical-based course in digital electronics, aimed at Advanced GNVQ, BTEC National and HNC/D students, degree foundation courses, and first year undergraduates.

The second thread is an explicit platform for EASY-PC Professional XM, and PULSAR, two computer aided design packages produced by Number One Systems Ltd of St Ives, Cambridgeshire in the UK. To this end it should appeal to hobbyists and professionals who require a step-by-step introduction to the use of that company's software.

Content

From an academic point of view this text should support most of the early content necessary for a digital electronics course: binary arithmetic, Boolean algebra and logic gates. The combinational logic section concentrates on adder circuits. Sequential logic covers the evolution of flip-flops, design and construction of both asynchronous and synchronous counters, ending with the design and construction of register circuits.

The third part considers logic circuit families and the place occupied by TTL and CMOS. The final part of the book deals with printed circuit technology, the process of converting a schematic circuit drawing into a printed circuit board layout and fabricating that layout. It finishes with a selection of practical ideas that can be drawn, 'captured' and ultimately developed into usable printed circuit boards.

Structure and requirements

The book is organized such that the reader can first consider and digest the necessary theory before tackling the step-by-step practicals at the end of each chapter. The student will either be able to use this text in a classroom/workshop environment or at home.

Some computer knowledge is necessary. It is assumed that the reader will have access to a PC, be familiar with DOS operations and be able to use a simple text editor such as EDIT. The computer used should be able to run Windows applications as well as EASY-PC Professional XM and PULSAR at a reasonable speed using the extended memory mode. The software is Windows '95 and NT compatible. Full installation and configuration details will be on the disk supplied, in a READ.ME file.

Summary

This book has a considerable practical content and has an additional attraction in its close association with computer aided design. There is a 'blow-by-blow' guide to the use of EASY-PC Professional XM, a schematic drawing and printed circuit board (PCB) design computer package. The guide also conducts the reader through logic circuit simulation using PULSAR. Chapters on PCB physics and PCB production techniques make the book unique, and with its host of project ideas make it an ideal companion for the integrative assignment and common skills components required by BTEC and the key skills demanded by GNVQ.

The principal aim of this book is to provide a straightforward approach to the understanding of digital electronics. It is intended for those who have had little or no experience of the subject.

It is particularly suited to the BTEC National Certificate microelectronic and digital electronic units and the Higher National Certificate units on digital electronic system. Students preparing for degree level courses in electronic or computer science based subjects will find the text particularly helpful for foundation level studies.

Those who prefer the 'Teach-In' approach or would rather experiment with some simple circuits should find the book's final chapters on printed circuit board production and project ideas especially useful.

An instructor's manual containing solutions and additional suggestions will be available to lecturers.

R.I. Monk, Anglia Polytechnic University

Software installation and configuration

This section gives an overview of the software supplied with this book. The minimum necessary level of equipment is described, though more powerful equipment will of course improve performance. Typical and special installations are then discussed, with any precautions required.

Please note that the software provided with this book has restricted functionality compared with the full commercial versions. It will allow a maximum of 10 components to be used; only the default library DEMOLIB may be used; component creation is disabled; and some output options have been omitted. Generators cannot be created or edited within PULSAR.

Overview of EASY-PC Professional XM and PULSAR restricted versions

EASY-PC Professional XM is powerful, yet easy to use, third generation software for the production of quality printed circuit board (PCB) layouts and schematic circuit diagrams. You can draft circuit diagrams beyond A1 in size. Designs can be simulated in the PULSAR digital simulator without leaving EASY-PC Professional XM.

Having verified a design, you switch into PCB layout mode which places your components and connections ready on the screen. You position the components within a board outline and route the connections. The 'rats nest' of unrouted connections can be flexibly displayed or hidden. EASY-PC Professional XM provides net optimization to help route tracks and design rule checking to validate routes against the schematic check clearance specifications. The fully functioning version integrates seamlessly with a high performance shape based autorouter, available as an optional extra.

PULSAR accepts net-lists from EASY-PC Professional XM, using them as a basis for simulating the timing and logic states of the circuit. The display will clearly show propagation delays, warn of glitches, and highlight indeterminate states. The cursors allow accurate timing measurements, and report timing 'snapshots'. Customized state generators allow circuits to be stimulated in ingenious patterns, although these cannot be changed in the restricted version.

Minimum hardware configuration

EASY-PC Professional XM requires a PC with a 386DX or better processor running DOS 4.01 or later, with a VGA graphics adapter and monitor (preferably colour). 6 Mbytes free hard disk space and a minimum of 4 Mbytes of RAM are necessary. PULSAR requires a 386SX or better processor, EGA or better graphics, 2 Mbytes of disk space, and 1 Mbyte of memory. An appropriate mouse or tracker ball driver must be installed prior to running either program.

Both programs provide output to 9 or 24 pin Epson or IBM compatible dot matrix printers, and HP LaserJet compatible laser and inkjet printers. EASY-PC Professional XM also outputs to HPGL pen plotters. PULSAR also outputs in GEM .IMG format. Output can be directed to your computer's parallel or serial ports, or to a file on disk.

The preferred computer for running this software is a 486DX processor, with 8 Mbytes of memory, and VGA colour graphics support, plus a mouse or tracker ball.

EASY-PC Professional XM keeps much of its program on the hard disk, loading it into memory as needed. This is very much faster if a disk caching program (such as Smartdrive) is used.

Installation

Installation is straightforward and involves no changes to your AUTOEXEC.BAT and CONFIG.SYS files, although you should ensure that the number following 'FILES =' in your CONFIG.SYS file is at least 30.

It is a good idea to make a backup copy of the distribution disks. You can use the DOS DISKCOPY or COPY commands to do this. After copying, place the distribution disks in a safe place, and install from your (write protected) working disks.

We strongly recommend that you accept the default path provided. If this is impossible, you may install the software in another location, but must then immediately reconfigure some settings. Check that you have at least 6 Mbytes of free space on your hard drive. If you have too little space, the Install program will terminate with an 'Insufficient Disc Space!' message.

To install the software, place disk 1 in your floppy disk drive, change to that drive (see your DOS manual if you are unsure of how to do this), and type INSTALL `RETURN`. When the Install program runs, follow the instructions given on your display.

Where a menu choice is offered you can use the up and down arrow keys to move the highlight bar over the menu entry that you wish to select. Pressing `RETURN` will select the highlighted entry. Where text input is required you can either accept the suggested response given by simply pressing `RETURN`, type in new information which will replace the suggested response, or edit the current response using the `LEFT ARROW`, `RIGHT ARROW`, `HOME`, `END`, `DELETE` and `BACKSPACE` keys. Pressing `RETURN` will accept the information currently on the response line.

Check your new directory for a file called README.TXT. If this file exists it will contain important last minute information that will not appear in this book. Use the DOS TYPE command to examine this file.

PULSAR is automatically installed in the same directory as EASY-PC Professional XM.

Installation in other drives

If you cannot install to the recommended default drive (normally 'C:'), you should, immediately upon running the program, change the library path (in the 'Preferences' screen – use the 'Settings' menu) and the default path ('Set path' from the 'File' menu). Both these paths must be set to the drive and directory in which the software has been installed. Any Windows PIF file must also be corrected to reflect the change by using the Windows PIF editor.

Installing under Windows 3.x

Following the normal DOS installation you may wish to install an icon to run your restricted version of EASY-PC Professional XM under Windows. EPCPROX.PIF, PROXM4.PIF, PROXM8.PIF, PROXM16.PIF and EPCP.ICO files have been included on disk 2, and should be copied to your working directory, or your Windows directory (as you wish).

Follow the usual Windows installation procedure, but indicate the PIF file as the executable file. Use EPCPROX.PIF or PROXM4.PIF for computers with 4 Mbytes of memory, PROXM8.PIF for computers with 8 Mbytes, and PROXM16.PIF for 16 Mbytes or more. The working directory should be your chosen EASY-PC Professional XM directory. If this is not the recommended default of C:\EPCPROX, you must also use the PIF editor to correct the information in the PIF file.

While the Properties box is still open, click on 'Change Icon...'. You will be warned that no icon is available, and offered those in Program Manager. Click on browse, and select the EASY-PC Professional XM icon file, EPCP.ICO. The icon will appear, and can then be accepted in the usual way. Full details of DOS application installation will be found in your Windows manual.

Note that it is still necessary for a mouse driver to have been installed under DOS.

Installing under Windows '95

The initial installation should be performed in DOS mode. As described for Windows 3.x, the appropriate PIF and icon files must be copied to the working directory, but must then be renamed EPCPROX.PIF and EPCPROX.ICO, respectively. Type 'EXIT' to return to Windows '95, then create a short cut to EPCPROX.EXE in the usual manner. Full details are given in your Windows '95 manual and help screens. Non-default installations will also need the path details in the Properties box to be corrected.

Running EASY-PC Professional XM restricted version

Once you have installed EASY-PC Professional XM, running the program is simple. Just change into the directory containing EASY-PC Professional XM and type EPCPROX `RETURN`. The program will load and run, showing the title screen. With any version of Windows, just double click on the appropriate icon as usual.

More guidance is given as it is needed while you progress through this book. Remember that you need to be in the right directory before running EASY-PC Professional XM. (Including it in the path statement won't work – there are overlay files that wouldn't be found.)

Unrestricted versions of the software

Full information on the unrestricted versions of the software, and other useful products, is available from:

Number One Systems
Harding Way
St Ives
Huntingdon
Cambridgeshire PE 17 4WR
UK

Phone: +44 (0)1480 461778
Fax: +44 (0)1480 494042
e-mail: sales@numberone.com
Web: http://www.numberone.com

Number One Systems
126 Smith Creek Drive
Los Gatos,
CA 95030
USA

Phone: (408) 395 0249
Fax: (408) 395 0249

EASY-PC, EASY-PC Professional, EASY-PC Professional XM, EASY-LINK, ANALYSER III, ANALYSER III Professional, FILTECH, FILTECH Professional, LAYAN, MultiRouter, PULSAR, PULSAR Professional, Z-Match and Z-Match Professional are trade marks of Number One Systems Ltd. All other trade marks acknowledged. Specifications are subject to change without notice.

Part 1

1 Number systems and computer arithmetic

Introduction

Historically, counting operations have evolved from quite strange sequences into the decimal number system that is virtually universal today. The number of fingers a human being possesses may have had a great deal to do with why we count in tens. Digital systems, especially computers, have two discernible states – HIGH and LOW, OFF and ON etc. – so counting systems that use just two characters are equally important to electronic engineers. This chapter examines four different number systems, then looks at a variety of arithmetic techniques that are used in digital electronics.

Generally you will find that number systems in written texts, computer programs and the like default to the decimal system. In circumstances when mixed number systems are being used, the system or number base is indicated by a subscript character at the end of the number string. For example, 1234 decimal is written as 1234_{10}. Omitting the subscript 10 could cause confusion, because the magnitude of 1234_{10} is nothing like that of 1234_8 (octal) or even 1234_{16} (hexadecimal). We will see to what extent this is so later on in the chapter.

Number systems

Decimal system

The decimal system (denary) counts to a base of 10. It has a radix of 10, using 10 digits, 0 to 9 inclusive.

Take 5362.4_{10} for example. This string of numbers represents 5 thousands, 3 hundreds, 6 tens, 2 units and 4 tenths. Positionally we say that the 5 occupies the most significant position in the string and the 4 the least significant. Alternatively we can say that the 5 carries the most weight and the 4 the least weight. The number can be represented in terms of its decimal weights thus:

$$5000 + 300 + 60 + 2 + 0.4$$

or

$$(5 \times 1000) + (3 \times 100) + (6 \times 10) + (2 \times 1) + (4 \times 0.1)$$

or

$$5 \times 10^3 + 3 \times 10^2 + 6 \times 10^1 + 2 \times 10^0 + 4 \times 10^{-1} = 5362.4_{10}$$

Binary system

The binary system counts to a base of 2. It has a radix of 2, and 2 digits 0 and 1. These are known as BInary digiTs, or bits.

Take 101101.01_2 for example. Using the same strategy as for denary numbers, the example can be explained in terms of each bit weight. The magnitude of the number can be expressed more satisfactorily in terms of its decimal value. However, instead of powers of ten, the binary system has column weights to the power of 2:

$$2^5 \quad 2^4 \quad 2^3 \quad 2^2 \quad 2^1 \quad 2^0 \quad 2^{-1} \quad 2^{-2}$$

The columns have the decimal weights of:

$$32 \quad 16 \quad 8 \quad 4 \quad 2 \quad 1 \quad \tfrac{1}{2} \quad \tfrac{1}{4}$$

our example can then be expressed as

$$1 \times 2^5 + 0 \times 2^4 + 1 \times 2^3 + 1 \times 2^2 + 0 \times 2^1 + 1 \times 2^0 + 0 \times 2^{-1} + 1 \times 2^{-2}$$

and represented as a decimal magnitude thus:

$$32 + 0 + 8 + 4 + 0 + 1 + 0 + 0.25$$

$$= 45.25_{10}$$

Using five bits, the binary count is simply

Binary	Decimal	Binary	Decimal
0 0000	0	1 0000	16
0 0001	1	1 0001	17
0 0010	2	1 0010	18
0 0011	3	1 0011	19
0 0100	4	1 0100	20
0 0101	5	1 0101	21
0 0110	6	1 0110	22
0 0111	7	1 0111	23
0 1000	8	1 1000	24
0 1001	9	1 1001	25
0 1010	10	1 1010	26
0 1011	11	1 1011	27
0 1100	12	1 1100	28
0 1101	13	1 1101	29
0 1110	14	1 1110	30
0 1111	15	1 1111	31

The binary system is simple and reliable and is all that is required by a two-state electronic system. Unfortunately human beings find these strings of binary digits cumbersome and rather difficult to manage. The next two number systems help to overcome this problem.

Octal system

The octal system counts to a base of 8. It has a radix of 8 and digits 0 to 7 inclusive and is used to represent groups of three binary digits.

Take 172.4_8 for example. Just as in the binary system, each digit can be expressed as a decimal weight, this time using column values to powers of eight:

$$8^3 \quad 8^2 \quad 8^1 \quad 8^0 \quad 8^{-1}$$

which have decimal weights of:

$$512 \quad 64 \quad 8 \quad 1 \quad \tfrac{1}{8}$$

Our example can be represented as:

$$1 \times 8^2 + 7 \times 8^1 + 2 \times 8^0 + 4 \times 8^{-1}$$
$$= 64 + 56 + 2 + 0.5 = 122.5_{10}$$

Hexadecimal system

The hexadecimal system counts to a base of 16. Sixteen characters are required, and this system uses the numerals 0 to 9 and the letters A to F for the extra six states from 10_{10} to 15_{10}. The hexadecimal system best represents groups of four binary digits.

Take $1AF.8_{16}$ for example. Each digit has a column value with a decimal weight to the power of 16:

$$16^3 \quad 16^2 \quad 16^1 \quad 16^0 \quad 16^{-1}$$

or which have decimal weights of:

$$4096 \quad 256 \quad 16 \quad 1 \quad \tfrac{1}{16}$$

Our example can be represented thus:

$$1 \times 16^2 + A \times 16^1 + F \times 16^0 + 8 \times 16^{-1}$$

or

$$1 \times 16^2 + 10 \times 16^1 + 15 \times 16^0 + 8 \times 16^{-1}$$
$$= 256 + 160 + 15 + \left(\tfrac{8}{16}\right)$$
$$= 431.5_{10}$$

Comparing the four systems:

Decimal	Binary	Octal	Binary	Hexadecimal
0	000 000	0	0 0000	0
1	000 001	1	0 0001	1
2	000 010	2	0 0010	2
3	000 011	3	0 0011	3
4	000 100	4	0 0100	4
5	000 101	5	0 0101	5
6	000 110	6	0 0110	6
7	000 111	7	0 0111	7
8	001 000	10	0 1000	8
9	001 001	11	0 1001	9
10	001 010	12	0 1010	A
11	001 011	13	0 1011	B
12	001 100	14	0 1100	C
13	001 101	15	0 1101	D
14	001 110	16	0 1110	E
15	001 111	17	0 1111	F
16	010 000	20	1 0000	10
17	010 001	21	1 0001	11
18	010 010	22	1 0010	12
19	010 011	23	1 0011	13
20	010 100	24	1 0100	14
21	010 101	25	1 0101	15
22	010 110	26	1 0110	16
23	010 111	27	1 0111	17
24	011 000	30	1 1000	18
25	011 001	31	1 1001	19
26	011 010	32	1 1010	1A
27	011 011	33	1 1011	1B
28	011 100	34	1 1100	1C
29	011 101	35	1 1101	1D
30	011 110	36	1 1110	1E
31	011 111	37	1 1111	1F

The first 32 counts are illustrated here. Note the divisions between the groups of bits; just like the decimal system where digits are grouped in hundreds, binary digits are normally grouped in fours. This conforms to the hexadecimal shorthand. Alternatively, if octal systems are being used, it is best to group bits in threes.

Points to note

- The point separating the whole part of a number from its fractional part is known as the radix point. The decimal point is a radix point, its equivalent in the binary system is the binary point, etc.

- The length of a number in a particular system = Length of that number (with radix R) – Length of the equivalent decimal number.
  or, approximately,

  $$= 1 - \log 10_R$$

  e.g.

  $$\text{Binary}:\text{Decimal} = 1 - \log 10_2 = 3.33:1$$
  $$\text{Octal}:\text{Decimal} = 1 - \log 10_8 = 1.11:1$$
  $$\text{Hexadecimal}:\text{Decimal} = 1 - \log 10_{16} = 0.83:1$$
  $$\text{Octal}:\text{Binary} = 1 - \log 2_8$$
  $$= \text{a third of a binary number}$$
  $$\text{Hexadecimal}:\text{Binary} = 1 - \log 2_{16}$$
  $$= \text{a quarter of a binary number}$$

Number conversions

Converting to and from decimal numbers is merely a question of manipulating the decimal weights of the binary, octal and hexadecimal numbers. Conversion between binary, octal and hexadecimal is simply a case of mapping between groups of three or four bits.

Adding up the binary weightings is fine for small numbers, but when the binary numbers start getting large or fractional an alternative approach may be tried.

Binary to decimal

For example, let us convert 110110.11_2 to decimal.

Method 1.   Using the decimal weight method this can be solved thus:

```
1   1   0   1   1   0  .  1    1
32  16  0   4   2   0  .  0.5  0.25
```
$$= 32 + 16 + 4 + 2 + 0.5 + 0.25$$
$$= 54.75_{10}$$

Method 2.   Alternatively, you can work with pairs of bits and successively add their decimal values, starting with the most significant bits:

```
1   1   0   1   1   0  .  1   1
```
$$(2 \times 1) + 1 = 3 \text{ (note that } 11_2 = 3_{10})$$
$$(2 \times 3) + 0 = 6$$
$$(2 \times 6) + 1 = 13$$
$$(2 \times 13) + 1 = 27$$

$$(2 \times 27) + 0 = 54$$
$$[1 + (1/2)] = 3/2$$
$$(3/2) \times (1/2) = 3/4$$
$$= 54 + \tfrac{3}{4}$$
$$= 54.75_{10}$$

Decimal to binary

For example, let us convert 74_{10} to binary.

Method 1.   By inspecting the decimal weights we can tackle the problem this way:

64   32   16   8   4   2   1

$$74 \div 64 = 1 \text{ rem } 10$$
$$10 \div 32 = 0 \text{ rem } 10$$
$$10 \div 16 = 0 \text{ rem } 10$$
$$10 \div 8 = 1 \text{ rem } 2$$
$$2 \div 4 = 0 \text{ rem } 2$$
$$2 \div 2 = 1 \text{ rem } 0$$
$$0 \div 1 = 0$$

giving:

1   0   0   1   0   1   0

$$= 100\,1010_2$$

Method 2.   This involves continuous division by 2. Using the previous example, conversion becomes:

Decimal	Remainder
$74 \div 2 = 37$	0 (the least significant bit)
$37 \div 2 = 18$	1
$18 \div 2 = 9$	0
$9 \div 2 = 4$	1
$4 \div 2 = 2$	0
$2 \div 2 = 1$	0
1	1 (the most significant bit)

Answer $= 100\,1010_2$

This is perhaps the safer method.

When dealing with fractions, the conversion is done by continuous multiplication:

Example: convert 0.375_{10} to binary.

Decimal	Remainder fraction	Whole number
$0.375 \times 2 = 0.75$	0.75	0 (the most significant bit)
$0.75 \times 2 = 1.5$	0.50	1
$0.5 \times 2 = 1$	0	1

Answer $= 0.011_2$

Hexadecimal to binary and vice versa

For example, let us convert $3EF2_{16}$ to binary and then to octal.

Method. The simplest method is to replace each character with a 4-bit binary number:

```
3     E     F     2
0011  1110  1111  0010
```

$= 0011\,1110\,1111.0010_2$

Grouping the bits in threes we can write down the octal equivalent:

$001\,111\,101\,111.001_2 = 1757\,2_8$

The reader is now left to tackle number conversions for him or herself. There are a variety to choose from in the Exercises given at the end of this chapter.

Binary numbers

Binary arithmetic

The rules for binary arithmetic are no different from those we employ for decimal numbers. The mechanics are generally far simpler!

Addition rules

Augend A		Addend B		Sum S	Carry C
0	plus	0	=	0	0
0	plus	1	=	1	0
1	plus	0	=	1	0
1	plus	1	=	0	1

Example :   *Example :*

101	Augend	1110111
+1001	Addend	+0101101
1	Carry	111111
10110	Sum	10100100

Subtraction rules

Minuend A		Subtrahend B		Difference S	Borrow C
0	minus	0	=	0	0
0	minus	1	=	1	1
1	minus	0	=	1	0
1	minus	1	=	0	0

Example :   *Example :*

1110	Minuend	1101101
−1001	Subtrahend	−0110110
	Borrow	1111
0101	Difference	0110111

Multiplication rules

These are very simple:

$$0 \times 0 = 0, 0 \times 1 = 0, 1 \times 0 = 0, 1 \times 1 = 1$$

Example :   *Example :*

11111	Multiplicand	0.111
×10010	Multiplier	×0.101
0	Partial products	0.0111
11111		0.000111
00		
11111		0.100011
1000101110	Product	

Division

Division is performed by repeated subtraction, just as with base 10 numbers. It's easier though, as the only multiples are 0 and 1. When you carry the next figure down from the dividend to add to the end of the previous remainder, it's immediately obvious whether you can subtract the divisor. Let's look at some examples to see how this works out in practice.

Let us divide 110010_2 by 101_2:

$$
\begin{array}{r}
00\,1010 \text{ Quotient} \\
\text{Divisor } 101\,\big|\,11\,0010 \text{ Dividend} \\
101 \\
\hline
101 \\
101 \\
\hline
\end{array}
$$

The divisor will go into the first three most significant bits once, with a remainder 1_2. Add the next bit to the remainder – giving 10_2. This is less than the divisor so will return a zero, then add the next bit to the remainder 10_2 giving 101_2. The divisor goes into this exactly once, giving an answer 1010_2 with no remainder.

A quick conversion of the problem into decimal will check whether the answer is correct:

$$50_{10} \div 5_{10} = 10_{10}.$$

As a second example, let us divide 100011_2 by 110_2:

$$
\begin{array}{r}
0\,0011.0010\,101 \text{ etc.} \\
110\,\big|\,1\,0011.0000 \\
110 \\
\hline
111 \\
110 \\
\hline
10.00 \\
1.10 \\
\hline
1000, \text{ etc.}
\end{array}
$$

As an exercise, convert the above to decimal and check result $(19_{10} - 6_{10} = 3.1667_{10})$.

Representation of negative numbers

A computer must be able to handle both positive and negative quantities, i.e. it must be able to represent the sign of a number as well as the number itself.

Sign and magnitude

In this notation the first, or most significant, bit indicates the sign. Convention has it that a '0' represents a positive number and '1' a negative number. Thus $0101_2 = +5$ and $1011_2 = -3$.

Complements

For a binary number N_2, we can define the true complement of N_2 as $2n - N_2$ and the digital complement of N_2 as $[2n - N_2] - 2 - m$; where n is the number of bits in the integer part and m is the number of bits in the fractional part.

Practical methods of forming the binary complements

- Digital complement: the rule is to change 1s for 0s and 0s for 1s in the binary number.
- True complement: add a 1 to the right-hand, or least significant, bit in the digital complement.

Examples:

Binary number (N_2)	Digital complement	True complement
1011	0100	0101
1010	0101	0110
1001.01	0110.10	0110.11
0.11010	1.00101	1.00110
0.0101	1.1010	1.1011

Binary complement subtraction

Earlier, the method was shown thus

$$25_{10} - 12_{10} = 11001_2 - 01100_2 = 01101_2$$

Using the same example we will investigate two methods of subtraction that involve binary addition. Note that in all instances both the minuend and the subtrahend must be the same length.

Digital complement subtraction

The subtrahend is changed to the digital complement 10011_2 and is added to the minuend, giving 101100_2. This is known as the 1s complement. The most significant bit is an overflow and indicates that the number is positive. This is wrapped around and added back to the least significant bit, thus giving the required answer.

Example

$$9_{10} - 5_{10} = 01001_2 + 11010_2 = 100011_2$$

the most significant 1 signifies that the number is still a complement. Moving the 1 and inverting, we get $00100_2 = 4_{10}$.

Example

$$910 - (-5)10 = 01001_2 + 00101_2 = 01110_2$$

The most significant 0 signifies that the number is positive: we get $01110_2 = 14_{10}$.

True complement subtraction

In the first example the subtrahend is changed to true complement 10100_2 (which is $10011_2 + 000001_2$). This is then added to the minuend giving 101101_2, which is the required answer. This is known as the 2s complement. Again the most significant bit is an overflow, but merely indicates that the number is positive.

Example

$$9_{10} - 5_{10} = 0\,1001_2 + 1\,1011_2 = 0\,0100_2 = 4_{10}$$

Example

$$9_{10} - (-5)_{10} = 0\,1001_2 + 0\,0101_2 = 0\,1110_2 = 14_{10}$$

Number Representation in Digital Circuits

Floating point representation

So far we have seen that a collection of bits can represent the whole, or integer, part of a number, with a similar set representing the fractional part. Both these are separated by the radix or binary point which is at a fixed position in the number. This method is known as *fixed point representation*. The fixed point method makes inefficient use of a computer's memory because, when saving an integer, all the bits reserved for the fractional part would be zero and thus be wasted as they convey no information. Conversely, the integer part of an entirely fractional number would be wasted also.

Scientific processing demands that both very large and very small numbers have to be manipulated with these numbers stored to a high degree of precision. One way of achieving this is by means of *floating point arithmetic*.

For example, the fixed point decimal number 1234.5678 can be expressed as $1.234\,567\,8 \times 10^3$ and 0.0076 can be written as 7.6×10^{-3}. In this amended form the first part of the number is referred to as the mantissa and the second part as the exponent. Here we have the engineering or scientific form of a number. Taking this one step further, the mantissa can be manipulated so that it is always expressed as a fraction, the first number becoming $0.123\,456\,78 \times 10^4$ and the second 0.76×10^{-2}. The decimal point has appeared to have floated to a position in the number where the most significant digit occupies the $\frac{1}{10}$ position. This method is widely used to represent numbers in electronic circuits.

In computing language a collection of 8 bits is referred to as a byte. Two bytes could therefore represent a floating point number, 1 byte denoting the mantissa or fractional part and the other the exponent. The bytes could be stored in sign and magnitude notation, 1s or 2s complement or a combination of all three – but with one sign bit, the seven remaining bits in the mantissa could only represent a number in the range $\pm 32\,768$ and the exponent as a power of 10, in the range ± 127. An impressive range, but it can be shown that, in spite of this, we only have an accuracy of two decimal places. Precision can be increased by doubling the number of bytes (11 bits would be required for three decimal place accuracy). Rounding errors may be kept to a minimum by forming a mantissa as near to the value 1 as possible, but where accuracy is important integer arithmetic is usually preferred.

Binary coded decimal

Another method of representing decimal numbers in an electronic system is by binary coded decimal (BCD). Each number in the range 0 to 9 can be represented by its binary equivalent, and therefore groups of 4 bits can represent any decimal number.

There are several BCD systems, of which the 8421-BCD is the most popular. In the 8421 system the BCD number is given the pure binary weighting of each bit. No binary numbers above 1001_2 are used. Another system is the less popular 2421-BCD; here the most significant bit has a weighting of 2 and the others 421, as before. This then only differs in the representation of 8_{10} and 9_{10}.

Examples:

Decimal	8421	2421
0	0 000	0 000
1	0 001	0 001
2	0 010	0 010
3	0 011	0 011
4	0 100	0 100
5	0 101	0 101
6	0 110	0 110
7	0 111	0 111
8	1 000	1 110
9	1 001	1 111

Examples:

Decimal	8421-BCD	2421-BCD
22	0010 0010	0010 0010
98	1001 1000	1111 1110
37	0011 0111	0011 0111
144	0001 0100 0100	0001 0100 0100

Exercises

1. Convert the following binary numbers into their (unsigned) decimal equivalents:
   (a) $1111\ 0001_2$ (b) $1010\ 1010_2$
   (c) $1100\ 1100_2$ (d) $1000\ 1101_2$

2. Convert the following decimal numbers into binary form by using the 2s method:
   (a) 240_{10} (b) 87_{10} (c) 163_{10} (d) 149_{10}

3. Convert the following octal numbers into their decimal equivalents:
   (a) 225_8 (b) 101_8 (c) 745_8 (d) 176_8

4. Convert the following decimal numbers into their octal equivalents:
   (a) 284_{10} (b) 659_{10} (c) 100_{10} (d) 325_{10}

5. Convert the following binary numbers into their octal equivalents:
   (a) $1\,1001\,0111_2$ (b) $1110\,1110_2$
   (c) $1\,0001\,1010_2$ (d) $1\,0111\,0111_2$

6. Convert the following octal numbers into their binary equivalents:
   (a) 361_8 (b) 746_8 (c) 122_8 (d) 37_8

7. Convert the following hexadecimal numbers into their (unsigned) decimal equivalents:
   (a) $E5_{16}$ (b) $2C_{16}$ (c) 98_{16} (d) $F1_{16}$

8. Convert the following decimal values into their hexadecimal equivalents:
   (a) 54_{10} (b) 200_{10} (c) 91_{10} (d) 238_{10}

9. Convert the following binary numbers into their hexadecimal equivalents:
   (a) $1101\,0111_2$ (b) $1110\,1010_2$
   (c) $1000\,1011_2$ (d) $1010\,0101_2$

10. Convert the following hexadecimal numbers into their binary equivalents:
    (a) 37_{16} (b) ED_{16} (c) $9F_{16}$ (d) $A2_{16}$

11. Using the rules for binary addition, evaluate the following:
    (a) $1010_2 + 111_2$ (b) $1111_2 + 1001_2$
    (c) $1011_2 + 1010_2$ (d) $1110_2 + 110_2$

12. Using the rules for octal addition, evaluate the following:
    (a) $33_8 + 27_8$ (b) $76_8 + 54_8$
    (c) $12_8 + 7_8$ (d) $67_8 + 17_8$

13. Using the rules for hexadecimal addition, evaluate the following:
    (a) $3E_{16} + 2C_{16}$ (b) $A8_{16} + 9B_{16}$
    (c) $6F_{16} + 29_{16}$ (d) $BB_{16} + 1A_{16}$

14. Using the rules for binary subtraction, evaluate the following:
    (a) $1111_2 - 101_2$ (b) $1011_2 - 100_2$
    (c) $1110_2 - 101_2$ (d) $1010_2 - 1000_2$

15. Using the rules for octal subtraction, evaluate the following:
    (a) $62_8 - 35_8$ (b) $71_8 - 16_8$
    (c) $24_8 - 15_8$ (d) $76_8 - 07_8$

16. Using the rules for hexadecimal subtraction, evaluate the following:
    (a) $A5_{16} - 68_{16}$ (b) $F2_{16} - BC_{16}$
    (c) $9F_{16} - 3D_{16}$ (d) $81_{16} - 1B_{16}$

17. Determine the 2s complement of the following binary numbers:
    (a) $1001\,1100_2$ (b) $1111\,0000_2$
    (c) $1010\,1010_2$ (d) $1101\,1100_2$

Continued on p. 20

Exercises (*Continued*)

18. Show how the following decimal numbers may be represented by using 8-bit 2s complement values:
    (a) -238_{10} (b) -2_{10} (c) -65_{10} (d) -93_{10}

19. Convert the following decimal numbers into their equivalent hexadecimal values:
    (a) -20_{10} (b) -115_{10} (c) -36_{10} (d) -99_{10}

20. Evaluate the following by using the 2s complement method for subtraction:
    (a) $1111\,0111_2 - 1\,1011_2$ (b) $1000\,0000_2 - 1\,1111_2$
    (c) $1011\,0110_2 - 1000\,0001_2$ (d) $1101\,0010_2 - 1000\,0111_2$

2 Logic functions and Boolean algebra

Introduction

As well as being able to perform arithmetic, digital circuits must also be able to make decisions according to a set of rules known as the *laws of logic*. The laws of logic were 'invented' in the 1850s by an English mathematician who attempted to describe the operation of the entire universe in terms of statements which were either true or false. His statements could be written down mathematically, and what emerged is now known as *Boolean algebra*.

In modern electronics these simple true/false arguments are best represented as high and low voltages, on or off states, or in binary as logic 0 and logic 1. PULSAR illustrates this nicely (Figure 2.1). The top trace shows a single variable changing from a logic 0 to a logic 1 after 150 µs. The second trace shows a serial string of binary data that changes level every 50 µs and which represents the binary number 100100101_2.

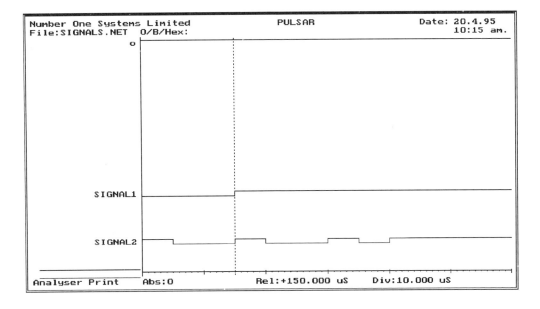

Figure 2.1

Logic gates

A device which is capable of performing a Boolean operation is known as a logic gate. The term gate is useful because it suggests that it can be controlled in some way to allow certain signals to pass through whilst inhibiting others. All modern digital electronic circuits are built up from these simple gates.

Variables

Digital signals or variables can be indicated by letters of the alphabet. Each variable will have just two values or states: 0 or 1. If it is NOT in one state, it must be in the other. Convention is to think of an open switch contact as representing logic 0 and a closed contact representing logic 1.

Consider variable 'A':

If $A = 0$, NOT $A = 1$ or $-A = 1$ or $NA = 1$ or $\overline{A} = 1$ or $/A = 1$

If $A = 1$, NOT $A = 0$ or $-A = 0$ or $NA = 0$ or $\overline{A} = 0$ or $/A = 0$

Representing the NOT function does present a problem and you will no doubt encounter a variety of different forms. In these two expressions a minus sign, the letter N, a forward stroke / and an overbar have been used to describe the NOT condition. The asterisk (*) symbol is also used in many texts.

We can discount the minus sign because of its conflict with the arithmetic function. Pure Boolean algebra uses the overbar, but this is rather difficult to type. This text will use / for single and grouped variables and an overbar wherever possible.

EASY-PC and other computer-aided design packages do not allow the use of non-alphanumeric characters in their variable names, so the letter N is used to represent NOT.

The AND function

This is best described by the simple electronic circuit shown in Figure 2.2. For the lamp to light up, both switch A and switch B have to be turned on. But four possible switch combinations are possible and the different arrangements can be described in a

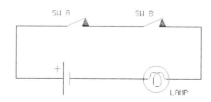

Figure 2.2

truth-table:

Switch *A*	Switch *B*	Lamp state
Open	Open	Off
Open	Closed	Off
Closed	Open	Off
Closed	Closed	On

Representing switches A and B by a 0 when open and a 1 when closed and the lamp or output as 0 for off and 1 for on, we can reconstruct the truth-table thus:

A	*B*		Output *Q*
0	0	NOT *A* and NOT *B*	0
0	1	NOT *A* and *B*	0
1	0	*A* and NOT *B*	0
1	1	*A* and *B*	1

All the states of switch A and switch B are examined. The fourth state is the only true state, indicated by the solitary 1.

The Boolean function for *A* AND *B* is written $A \cdot B$, the dot being the AND operator. As conventional with products in algebra, the dot may be omitted, and the expression written as AB.

The OR function

The electronic switch circuit now becomes as shown in Figure 2.3. For the lamp to turn on, either switch A or switch B has to be closed. Representing all possible arrangements of the switches we get a truth-table:

Switch *A*	Switch *B*	Lamp state
Open	Open	Off
Open	Closed	On
Closed	Open	On
Closed	Closed	On

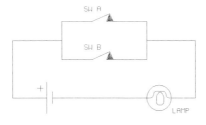

Figure 2.3

Representing switches A and B by a 0 when open and a 1 when closed and the lamp or output as 0 for off and 1 for on, we can reconstruct the truth-table thus:

A	B		Output Q
0	0	NOT A and NOT B	0
0	1	NOT A and B	1
1	0	A and NOT B	1
1	1	A and B	1

Here three states are true: NOT A and B, A and NOT B and A and B.

The Boolean function for A OR B is written $A + B$, the plus sign being the OR operator.

Multiple variables

We are not restricted to only two input variables. Three, four, five and more may be used, but the solution of Boolean equations then becomes more difficult.

Let us consider the two examples shown in Figure 2.4:

$$Q = A + B + C$$

$$Q = A \cdot B \cdot C, \text{ or } ABC$$

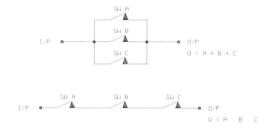

Figure 2.4

Expressing these circuits as truth-tables we get:

A	B	C	$A + B + C$	$A \cdot B \cdot C$
0	0	0	0	0
0	0	1	1	0
0	1	0	1	0
0	1	1	1	0
1	0	0	1	0
1	0	1	1	0
1	1	0	1	0
1	1	1	1	1

As we have seen, the AND operator is frequently omitted when coupling several variables together. AND is also assumed to operate on brackets if the OR operator is not used. Also, a bar that couples several variables can be considered as bracketing those variables together. For example $A \cdot B \cdot C$ is the same as ABC and $A \cdot (B + C)$ is the same as $A(B + C)$. $\overline{A + C}$ is also the same as $\overline{(A + C)}$.

Simplification

Since it is possible to construct electronic circuits from any given Boolean expression, it becomes vital to reduce the expression to its simplest form first. The advantages of doing so are obvious: fewer components are needed and, therefore, the system is cheaper, more reliable, and requires a smaller circuit assembly.

The laws of Boolean algebra are given in Table 2.1. The list may appear frightening at first glance, but its contents are really quite straightforward. Proving that these rules are true can be done many ways – by using switch circuit analogy, truth-tables or by applying Boolean laws that have already justified. The first eight laws will be proven using switch circuits.

Table 2.1 *The laws of Boolean algebra*

	Law	Property type
(a)	$A \cdot 0 = 0$	
(b)	$A \cdot 1 = A$	
(c)	$A \cdot A = A$	Idempotent
(d)	$A \cdot /A = 0$	Complement
(e)	$A + 0 = A$	
(f)	$A + 1 = 1$	
(g)	$A + A = A$	Idempotent
(h)	$A + /A = 1$	Complement
(i)	$//A = A$	Double complement
(j)	$A \cdot B = B \cdot A$	Commutative
(k)	$A + B = B + A$	Commutative
(l)	$(A \cdot B) \cdot C = A \cdot (B \cdot C)$	Associative
(m)	$(A + B) + C = A + (B + C)$	Associative
(n)	$A(B + C) = A \cdot B + A \cdot C$	Distributive
(o)	$A + (B \cdot C) = (A + B) \cdot (A + C)$	Distributive
(p)	$A \cdot (A + B) = A$	Absorption law
(q)	$A + A \cdot B = A$	Absorption law
(r)	$A + /A \cdot B = A + B$	Absorption law
(s)	$/(A \cdot B) = /A + /B$	De Morgan's theorem
(t)	$/(A + B) = /A \cdot /B$	De Morgan's theorem

The AND laws

(a) $A \cdot 0 = 0$
   The circuit is always open, no matter what position switch A is in.
(b) $A \cdot 1 = A$
   The 1 is a short circuit, so the output only reflects the state of switch A.

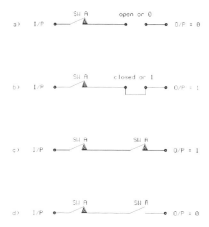

Figure 2.5

(c) $A \cdot A = A$

Both switches are 'ganged' together and are therefore both on together and off together. The rule implies that any number of A switches As connected in series will simplify to a single switch.

(d) $A \cdot \overline{A} = 0$

Both switches are 'ganged' together and one is always off when the other is on. The circuit is therefore always open.

The OR laws

(e) $A + 0 = A$

The logic 0 is an open circuit, and hence the circuit only sees switch A.

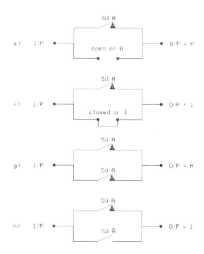

Figure 2.6

(f) $A + 1 = 1$
  The logic 1 is a short circuit, so the switch will have no effect. The circuit is always closed.

(g) $A + A = A$
  Both switches are 'ganged' together and are therefore both on together and off together. The rule implies that any number of A switches connected in parallel will simplify to a single switch.

(h) $A + \overline{A} = 1$
  Both switches are 'ganged' together and one is always off when the other is on. The circuit is therefore always closed.

Double complement

(i) $\overline{\overline{A}} \equiv$ *not not A,* which simply reverts to A
  The double NOT is most likely to occur during simplification of larger expressions.

Commutative property

If variables show this property the order in which they appear within one logic expression is unimportant.

(j) $A \cdot B \cdot C = C \cdot B \cdot A = B \cdot A \cdot C \ldots$

(k) $A + B + C = C + B + A = B + A + C \ldots$

Associative laws

These laws allow variables contained within one logic or bracketed expression to be combined or separated.

(l) $(A \cdot B) \cdot C = A \cdot (B \cdot C)$

(m) $(A + B) + C = A + (B + C)$

Note: The AND operator is normally assumed before and after brackets, so equation (l) may well look like $(A \cdot B)C = A(B \cdot C)$.

Distributive property

This allows logic variables within one logic or bracketed expression to be expanded out or contracted. This is similar to ordinary algebraic manipulations.

(n) $A + (B \cdot C) = (A + B) \cdot (A + C)$

(o) $A \cdot (B + C) = (A \cdot B) + (A \cdot C)$

The rules for double inversion, the commutative property, and the associative and distributed laws may best be proven by translating the equations into truth-tables and comparing the output patterns. Having established their validity, they will be used to prove or justify the remaining laws.

Absorption laws

These three rules are very useful tools for expression minimization. They may be used to eliminate, cancel or absorb redundant variables in large Boolean equations.

(p) $A(A + B) = A$

which can be written as $A \cdot (A + B) = A$.
    We may prove this by using truth-tables:

(1) A	(2) B	(3) $A + B$	(4) $A \cdot (A + B)$
0	0	0	0
0	1	1	0
1	0	1	1
1	1	1	1

Columns (1) and (2) list every state of A and B. If we OR A and B together we get column (3), which was shown earlier on this chapter. Now if we AND A to column (3) to get the expression $A \cdot (A + B)$, we get column (4). Comparing the result with column (1) we see that they are the same, so it is safe to say that $A \cdot (A + B) = A$.

(q) $A + A \cdot B = A$

which can be written as $A + (A \cdot B) = A$.
    This can be proven the same way:

(1) A	(2) B	(3) $A \cdot B$	(4) $A + (A \cdot B)$
0	0	0	0
0	1	0	0
1	0	0	1
1	1	1	1

Column (1) is identical to column (4).
    Alternatively, using Boolean Laws, we can say:

$$A + A \cdot B = A \cdot 1 + A \cdot B \qquad \text{law (b)}$$
$$= A \cdot (1 + B) \qquad \text{law (o)}$$
$$= A \cdot 1 \qquad \text{law (f)}$$
$$= A \qquad \text{law (b)}$$

(r) $A + \overline{A} \cdot B = A + B$

Proof:

$$A + \overline{A} \cdot B = A \cdot (A + \overline{A}) + \overline{A} \cdot B \qquad \text{law (h)}$$
$$= A \cdot A + A \cdot \overline{A} + \overline{A} \cdot B \qquad \text{law (n)}$$
$$= A + \overline{A} \cdot (A + B) \qquad \text{law (n)}$$

Truth-table solution:

(1) A	(2) B	(3) NA	(4) NA · B	(5) A + (NA · B)	(6) A + B
0	0	1	0	0	0
0	1	1	1	1	1
1	0	0	0	1	1
1	1	0	0	1	1

In the table columns (3) and (4) provide intermediate results that give us column (5). Comparing column (5) with the table for *A* OR *B*, we see that they are the same.

De Morgan's theorem

This theorem allows a logic function to be represented in its other form. This is very useful for occasions when an AND expression is required in the form of an OR, or vice versa.

If $Q = A + C$, then $\overline{Q} = \overline{A + C}$ is equally as true. De Morgan's theorem, when applied to the latter expression, will alter its form to:

$$\overline{Q} = \overline{A} \cdot \overline{C}$$

so it is true to say that $\overline{A + C} = \overline{A} \cdot \overline{C}$.

In general, De Morgan's theorem can only be used when a function is a complement. The rule becomes *break the line change the sign*.

(s) $\overline{A \cdot B \cdot C \cdot D} = \overline{A} + \overline{B} + \overline{C} + \overline{D} + \ldots$

(t) $\overline{A + B + C + D+} = \overline{A} \cdot \overline{B} \cdot \overline{C} \cdot \overline{D} \cdot \ldots$

The proof of De Morgan's theorem may be obtained by using a truth-table. From the previous example, where $\overline{Q} = \overline{A + C} = \overline{A} \cdot \overline{C}$:

(1) A	(2) C	(3) A + C	(4) N(A + C)	(5) NA	(6) NC	(7) NA · NC
0	0	0	1	1	1	1
0	1	1	0	1	0	0
1	0	1	0	0	1	0
1	1	1	0	0	0	0

Columns (4) and (7) represent both sides of the equation and are clearly identical, so we have proved the relationship.

Example

Simplify the following expression by applying the laws of Boolean algebra:

$$Q = \overline{A + B} \cdot \overline{ABC} \cdot \overline{\overline{AC}}$$

Solution

The first section is expanded using De Morgan's theorem

$$Q = \overline{A} \cdot \overline{B} \cdot \overline{ABC} \cdot \overline{\overline{AC}}$$

$$= \overline{A} \cdot \overline{B} \cdot (\overline{A} + \overline{B} + \overline{C}) \cdot (\overline{\overline{A}} + \overline{C}) \qquad \text{(De Morgan)}$$

$$= ((\overline{A} \cdot \overline{B} \cdot \overline{A}) + (\overline{A} \cdot \overline{B} \cdot \overline{B}) + (\overline{A} \cdot \overline{B} \cdot \overline{C})) \cdot (A + \overline{C}) \qquad \text{laws(n) and(i)}$$

$$= ((\overline{A} \cdot \overline{B}) + (\overline{A} \cdot \overline{B}) + (\overline{A} \cdot \overline{B} \cdot \overline{C})) \cdot (A + \overline{C}) \qquad \text{law(c)}$$

moving the common factor outside the brackets to give 1s

$$= \overline{A} \cdot \overline{B} \cdot (1 + 1 + \overline{C}) \cdot (A + \overline{C})$$

$$= (\overline{A} \cdot \overline{B}) \cdot (A + \overline{C}) \qquad \text{laws(f) and(b)}$$

$$= (\overline{A} \cdot \overline{B} \cdot A) + (\overline{A} \cdot \overline{B} \cdot \overline{C}) \qquad \text{law(n)}$$

$$\text{Answer } = \overline{A} \cdot \overline{B} \cdot \overline{C} \qquad \text{law(d)}$$

Canonical forms of Boolean expression

There are two distinct, or canonical, forms of logic expressions, known as minterms and maxterms. *Minterms* are those terms which are ANDed together and represent each output state on the truth-table. *Maxterms* are those which are ORed together.

$$Q = (\overline{A} \cdot \overline{B} \cdot \overline{C}) + (\overline{A} \cdot \overline{B} \cdot C) + (A \cdot B \cdot \overline{C}) + (A \cdot B \cdot C)$$

The above equation is written entirely in minterms. It is also known as the *sum-of-products form.*

$$Q = (\overline{A} + \overline{B} + \overline{C}) \cdot (\overline{A} + \overline{B} + C) \cdot (A + B + \overline{C}) \cdot (A + B + C)$$

This second equation is written entirely in maxterms. It is also known as the *product-of-sums form.*

Karnaugh maps

Karnaugh devised a graphical method for representing Boolean equations on a two-dimensional diagram or map. The map is simply a set of squares, each square or cell corresponding to one minterm. Along the top and down one side of the map, the variables are listed so that they appear to change from cell to cell, row to row or column to column. Only one change is permitted across each cell boundary. Simplification then becomes a matter of grouping minterms in multiples of powers of 2, i.e. 2s, 4s, 8s, etc.

The standard grid for a *two-variable expression* is:

$$NA \cdot NB \quad NA \cdot B$$
$$A \cdot NB \quad \; A \cdot B$$

Each square holds just one minterm. Each row contains a unique form of the variable, as does each column. In the above example, the top row of the grid contains all NOT As, the bottom row all the As, with the left hand column containing the NOT Bs and the right one the Bs. The choice of either row or column is arbitrary, so long as only one variable changes state across each cell boundary. Consistency is the key of course, and this text will keep to one form unless declared otherwise.

Before a logic function can be plotted on a Karnaugh map it must be expressed in its sum-of-products form, with each product expression containing the same number of variables. A given equation may well have to be expanded to achieve this, as illustrated by the two examples given in the previous section on canonical forms. In both cases the variable C is redundant.

The Boolean expression $Q = \overline{A} \cdot B + A \cdot B$ would be mapped as:

$$0 \quad NA \cdot B$$
$$0 \quad \; A \cdot B$$

Writing 1s in the boxes instead of the minterms and indicating the variable names in headings for each row and column, the map becomes:

	NB	B
NA:	0	1
A:	0	1

By grouping these two elements together a simplified expression may be read directly from the grid:

	NB	B
NA:	0	1
A:	0	1

As the 1s occupy the B column, we deduce the expression $Q = \overline{A} \cdot B + A \cdot B = B$, as we know already from earlier Boolean algebra examples.

Rules for Karnaugh maps

- For N variables there are 2^N cells in the map.
- Place a 1 in each box corresponding to an element in a Boolean expression.
- Place a 0 in the empty cells unless a 'don't care' state is declared, in which case you put an x or d.
- Loop together 1s in groups of 2, 4, 8, 16, etc.; the larger the group the better.
- Loops may be adjacent vertically or horizontally, but not diagonally.
- Loops may overlap others to capture lone 1s.

- Any element which cannot be looped remains as a single element to be ORed with the rest of the expression.
- Only one variable changes across each boundary.
- Element positions on the map are fixed, but opposite boundaries are considered to be the same boundaries:

$A \cdot B$	$A \cdot NB$	$A \cdot B$	$A \cdot NB$
$NA \cdot B$	$NA \cdot NB$	$NA \cdot B$	$NA \cdot NB$
$A \cdot B$	$A \cdot NB$	$A \cdot B$	$A \cdot NB$
$NA \cdot B$	$NA \cdot NB$	$NA \cdot B$	$NA \cdot NB$

Three-variable maps

For a three-variable map, using variables A, B and C we need 2^3 or 8 cells:

$NA \cdot NB \cdot NC$	$NA \cdot B \cdot NC$	$A \cdot B \cdot NC$	$A \cdot NB \cdot NC$
$NA \cdot NB \cdot C$	$NA \cdot B \cdot C$	$A \cdot B \cdot C$	$A \cdot NB \cdot C$
$NA \cdot NB \cdot NC$	$NA \cdot B \cdot NC$	$A \cdot B \cdot NC$	$A \cdot NB \cdot NC$

Alternatively, we can translate these variables into their binary equivalents giving us a table like the following one:

000	010	110	100
001	011	111	101
000	010	110	100

or, more simply,

$C \backslash AB$	00	01	11	10
0:				
1:				

The states of A and B are displayed in the columns and the states of C are in the rows.

Example

Minimize the given Boolean expression using a Karnaugh map:

$$Q = A \cdot B \cdot \overline{C} + A \cdot \overline{B} \cdot C + \overline{A} \cdot \overline{B} \cdot C + \overline{A} \cdot B \cdot C$$

Solution

There are three variables, so we need to draw a three-variable map:

$C \backslash AB$	00	01	11	10
0:				
1:				

Place the variables on the map (remember, they must be minterms):

$C \backslash AB$	00	01	11	10
0:			$A \cdot B \cdot NC$	
1:	$NA \cdot NB \cdot C$	$NA \cdot B \cdot C$		$A \cdot NB \cdot C$

With experience, these get entered as 1s (see below). Now loop the groups together – the larger the better:

$C \backslash AB$	00	01	11	10
0:	0	0	1	0
1:	1	1	0	1

There are three groupings; two of two terms and one single term. Starting from the bottom left we have the first two. At the end of the row we have the second two; remember that the cell's right-hand boundary is shared with the extreme left-hand cell, so we have one minterm shared by two groups. Lastly, we have a solitary cell on the top row that does not fit into either of the other two groups. Remember that loopings cannot be diagonal. Translating each term we get, in the same orders $\overline{A} \cdot C$, $\overline{B} \cdot C$ and $A \cdot B \cdot \overline{C}$, which gives us our minimized expression:

$$Q = \overline{A} \cdot C + \overline{B} \cdot C + A \cdot B \cdot \overline{C}$$

Although the expression can be altered, it cannot be minimized further. Note that the complement of this expression is obtained by looping the '0's'.

Four-variable maps

For a four-variable map, using variables A, B, C, and D, we need 2^4, or 16, cells:

$NA \cdot NB \cdot NC \cdot ND$	$NA \cdot B \cdot NC \cdot ND$	$A \cdot B \cdot NC \cdot ND$	$A \cdot NB \cdot NC \cdot ND$
$NA \cdot NB \cdot NC \cdot D$	$NA \cdot B \cdot NC \cdot D$	$A \cdot B \cdot NC \cdot D$	$A \cdot NB \cdot NC \cdot D$
$NA \cdot NB \cdot C \cdot D$	$NA \cdot B \cdot C \cdot D$	$A \cdot B \cdot C \cdot D$	$A \cdot NB \cdot C \cdot D$
$NA \cdot NB \cdot C \cdot ND$	$NA \cdot B \cdot C \cdot ND$	$A \cdot B \cdot C \cdot ND$	$A \cdot NB \cdot C \cdot ND$

Alternatively, we can translate these variables into their binary equivalents, giving us a table like the following one:

0000	0100	1100	1000
0001	0101	1101	1001
0011	0111	1111	1011
0010	0110	1110	1010

or

$C \backslash AB$	00	01	11	10
00:				
01:				
11:				
10:				

Example

Minimize the given Boolean expression using a Karnaugh map:

$$Q = (\overline{A} \cdot \overline{B} \cdot \overline{C} \cdot \overline{D}) + (\overline{A} \cdot \overline{B} \cdot \overline{C} \cdot D) + (\overline{A} \cdot B \cdot \overline{C} \cdot \overline{D}) + (A \cdot \overline{B} \cdot C \cdot D)$$

$$+ (A \cdot B \cdot C \cdot D) + (\overline{A} \cdot B \cdot C \cdot \overline{D}) + (\overline{A} \cdot B \cdot \overline{C} \cdot D)$$

Solution

There are four variables, so we need to draw a four-variable map and place each minterm:

$C \backslash AB$	00	01	11	10
00 :	1	1	–	–
01 :	1	1	–	–
11 :	–	–	1	1
10 :	–	1	–	–

This gives us three groupings. The group of four at the top gives us $\overline{A} \cdot \overline{C}$. The 1 on the bottom row groups with the top right cell of the previous group and gives us $\overline{A} \cdot B \cdot \overline{D}$. Finally, the two 1s in the third row give us $A \cdot C \cdot D$. The expression minimizes to:

$$Q = \overline{A} \cdot \overline{C} + \overline{A} \cdot B \cdot \overline{D} + A \cdot C \cdot D$$

Maps with more than four variables

Five-variable maps are obtained by placing two four-variable maps side by side. One map for the fifth variable, E and the other for its complement. Larger maps are possible but become rather cumbersome and thus lose the principal advantage of Karnaugh maps, which is their simplicity. For expressions with more than five variables minimization techniques such as Quine–McClusky's algorithm (which is beyond the scope of this text) must be used.

Exercises

1. Simplify the following equations:

    (i) $Q = \overline{A + B} \cdot \overline{ABC} \cdot \overline{\overline{A}C}$

    (ii) $Q = ((A \cdot B) + (\overline{B} \cdot C)) + ((B \cdot \overline{C}) + (\overline{A} \cdot B))$

    (iii) $Q = ((A \cdot B) + (\overline{B} \cdot C)) \cdot ((A \cdot C) + (\overline{A} \cdot \overline{C}))$

    (iv) $Q = ((\overline{A} \cdot B) \cdot (A + \overline{C}))$

    (v) $Q = E \cdot (\overline{A + B + C})$

    (vi) $Q = (A + B) \cdot (\overline{A} \cdot \overline{B})$

    (vii) $Q = ABC + A(\overline{B} + \overline{C})$

(viii) $Q = \overline{A}CD + \overline{A}BD + ACD + ABD$

(ix) $Q = (X + \overline{Y}X) \cdot (X + YZ)$

(x) $Q = (A\overline{B} + \overline{A}B) \cdot (AB + \overline{A}\overline{B})$

(xi) $Q = A\overline{C} + ABC + AC$

(xii) $Q = (A + AB + ABC) \cdot (A + B + C)$

(xiii) $Q = (XY + ABC) \cdot (XY + \overline{A} + \overline{B} + \overline{C})$

(xiv) $Q = (\overline{A} \cdot \overline{B} \cdot C \cdot \overline{D}) + (\overline{A} \cdot \overline{B} \cdot C \cdot D) + (A \cdot \overline{B} \cdot C \cdot \overline{D}) + (A \cdot \overline{B} \cdot C \cdot D)$

(xv) $Q = BD + (\overline{B} + \overline{D}) \cdot C$

(xvi) $Q = E(A\overline{B} + C\overline{E}) + \overline{E}(\overline{A}B + \overline{C}E)$

2. Simplify the following equations by using (a) Boolean algebra and (b) a Karnaugh map

   (i) $Q = A \cdot \overline{B} + \overline{A} \cdot B \cdot + A \cdot B$

   (ii) $Q = A + \overline{A} \cdot B$

   (iii) $Q = \overline{A} \cdot B + A \cdot \overline{B} + \overline{A} \cdot \overline{B}$

   (iv) $Q = A \cdot \overline{B} + \overline{A} \cdot B$

   (v) $Q = \overline{A} \cdot B \cdot \overline{C} + A \cdot B \cdot \overline{C}$

   (vi) $Q = A \cdot B \cdot C + \overline{B} \cdot C + \overline{A} \cdot C$

   (vii) $Q = \overline{A} \cdot \overline{B} \cdot \overline{C} + A \cdot \overline{B} \cdot \overline{C}$

   (viii) $Q = \overline{A} \cdot B \cdot \overline{C} + A \cdot \overline{C} + \overline{A} \cdot \overline{B} \cdot C + A \cdot \overline{B} \cdot C$

   (ix) $Q = A \cdot B + \overline{A} \cdot B \cdot C + \overline{A} \cdot \overline{C} + \overline{A} \cdot \overline{B}$

3. Minimize the following Boolean expressions using the Karnaugh map technique:

   (i) $Q = \overline{A} \cdot B \cdot C + \overline{A} \cdot \overline{B} \cdot C + \overline{A} \cdot B \cdot \overline{C} + A \cdot B \cdot C + A \cdot B \cdot \overline{C}$

   (ii) $Q = A \cdot \overline{B} \cdot \overline{C} + A \cdot B \cdot \overline{C} + A \cdot \overline{B} \cdot C + A \cdot B \cdot C$

   (iii) $Q = A \cdot B \cdot \overline{C} + \overline{A} \cdot \overline{B} \cdot \overline{C} + A \cdot \overline{B} \cdot \overline{C} + \overline{A} \cdot B \cdot \overline{C}$

   (iv) $Q = A \cdot B \cdot C + B \cdot C + A \cdot \overline{B} \cdot C$

   (v) $Q = A \cdot \overline{C} + A \cdot B + B \cdot C$

   (vi) $Q = A \cdot \overline{B} + \overline{A} \cdot B \cdot C + \overline{A} \cdot B \cdot \overline{C}$

4. Minimize the following Boolean expressions using the Karnaugh map technique. Express the answer in both sums-of-products and products-of-sums forms. (*Hint*: Try using the complement.)

   (i) $Q = \overline{A} \cdot \overline{B} \cdot C \cdot \overline{D} + \overline{A} \cdot B \cdot C \cdot D + A \cdot B \cdot \overline{C} \cdot D + A \cdot \overline{B} \cdot \overline{C} \cdot D$
   $+ A \cdot \overline{B} \cdot C \cdot D + A \cdot B \cdot C \cdot D$

   (ii) $Q = A \cdot B \cdot \overline{C} \cdot \overline{D} + A \cdot \overline{B} \cdot C \cdot D + A \cdot B \cdot \overline{C} \cdot D + A \cdot \overline{B} \cdot \overline{C} \cdot D + A \cdot B \cdot C \cdot D$

   (iii) $Q = \overline{A} \cdot B \cdot C \cdot \overline{D} + A \cdot B \cdot C \cdot \overline{D} + \overline{A} \cdot B \cdot \overline{C} \cdot \overline{D} + \overline{A} \cdot B \cdot C \cdot D$
   $+ A \cdot B \cdot C \cdot D + A \cdot B \cdot \overline{C} \cdot \overline{D}$

Continued on p. 36

Exercises *(Continued)*

(iv) $Q = \overline{A} \cdot \overline{B} \cdot \overline{C} \cdot \overline{D} + A \cdot \overline{B} \cdot \overline{D} + \overline{B} \cdot C \cdot \overline{D} + A \cdot B \cdot \overline{C} \cdot D + A \cdot B \cdot C \cdot D$
$\quad + \overline{A} \cdot B \cdot D$

(v) $Q = \overline{A} \cdot B \cdot \overline{C} \cdot D + \overline{C} \cdot D + A \cdot B \cdot \overline{D} + \overline{A} \cdot B \cdot C \cdot \overline{D}$

(vi) $Q = A \cdot \overline{B} \cdot \overline{C} + A \cdot \overline{C} \cdot D + B \cdot C \cdot D + \overline{A} \cdot C \cdot D + \overline{A} \cdot B \cdot \overline{D}$

(vii) $Q = B \cdot \overline{C} \cdot \overline{D} + A \cdot \overline{C} \cdot \overline{D} + A \cdot B \cdot \overline{C} \cdot D + A \cdot \overline{B} \cdot \overline{C} \cdot D$

(viii) $Q = A \cdot \overline{C} \cdot \overline{D} + A \cdot \overline{B} \cdot D + A \cdot C \cdot \overline{D}$

(ix) $Q = C(A \cdot \overline{B} + A \cdot B \cdot \overline{D} + A \cdot \overline{B} \cdot D) + A \cdot \overline{C}$

Solutions

1. (i) $\overline{A} \cdot \overline{B} \cdot \overline{C}$ (v) $\overline{A} \cdot \overline{B} \cdot \overline{C} \cdot E$ (ix) X (xiii) $X \cdot Y$
    (ii) $\overline{A} \cdot B + C$ (vi) 0 (x) 0 (xiv) $\overline{B} \cdot C$
    (iii) $\overline{A} + \overline{C}$ (vii) A (xi) A (xv) $B \cdot D + C$
    (iv) $A + \overline{B} + C$ (viii) $D \cdot (B + C)$ (xii) A (xvi) $A \cdot \overline{B} \cdot E + \overline{A} \cdot B \cdot \overline{E}$

2. (i) $A + B$ (iv) Already minimized   (vii) $\overline{B} \cdot \overline{C}$ or $\overline{B + C}$
    (ii) $A + B$ (v) $\overline{C} \cdot B$ (viii) $A \cdot \overline{C} + B \cdot \overline{C} + \overline{B} \cdot C$
    (iii) $\overline{A} + \overline{B}$ (vi) C (ix) $\overline{A} + B$

3. (i) $B + \overline{A} \cdot C$ (iii) $\overline{C}$ (v) $A \cdot \overline{C} + B \cdot C$
    (ii) A (iv) $C(A + B)$ (vi) $Q = \overline{B}$

4. (i) $A \cdot D + B \cdot C \cdot D + \overline{A} \cdot \overline{B} \cdot C \cdot \overline{D}(A + C)$
    $(\overline{A} + D)(\overline{B} + D)(A + B + \overline{D})$

    (ii) $A \cdot D + A \cdot B \cdot \overline{C}$
    $A(B + D)(\overline{C} + D)$

    (iii) $B \cdot C + B \cdot \overline{D}$
    $B(C + \overline{D})$

    (iv) $\overline{B} \cdot \overline{D} + B \cdot D$
    $(B + \overline{D})(\overline{B} + D)$

    (v) $C \cdot \overline{D} + A \cdot B \cdot \overline{D} + B \cdot C \cdot \overline{D}$
    $(B + D)(\overline{C} + \overline{D})(A + C + D)$

    (vi) $\overline{A} \cdot \overline{B} \cdot \overline{C} + A \cdot \overline{C} \cdot D + \overline{B} \cdot \overline{C} \cdot \overline{D} + B \cdot C \cdot D$
    $(\overline{B} + D)(\overline{A} + B + \overline{C})(A + C + \overline{D})$

    (vii) $A \cdot \overline{C} + B \cdot \overline{C} \cdot \overline{D}$
    $C(A + B)(A + \overline{D})$

    (viii) $A \cdot \overline{B} + A \cdot \overline{D}$
    $A(\overline{B} + \overline{D})$

    (ix) $A \cdot \overline{B} + A \cdot \overline{C} + A \cdot \overline{D}$
    $A(\overline{B} + \overline{C} + \overline{D})$

3 Logic gates

Logic gates can be made from switches, magnetic or fluidic devices, but are mostly electronic devices. These become the electronic circuits which realize Boolean or logic functions. Many conventions are used to represent logic devices; the BS, or British Standard, and the MS, or American Military Standard, are just two. The symbols used by this text will be the MS type, principally because they are encountered universally and are also used by EASY-PC Professional XM, which you will be meeting shortly. The two versions are compared in Figure 3.1.

The term 'gate' most likely has an agricultural origin, since when a farmyard gate is open livestock may pass from one pen to another. Similarly, logic elements are regarded as gates since there is a flow of information through a system when a gate is open and none when a gate is closed.

The AND gate

The AND function is represented by the AND gate. In the three examples shown in Figure 3.2 the inputs are applied to the left-hand side and the outputs are taken from the right.

- For U1 the output at Y is $A \cdot B$.
- For U2 the output is $A \cdot B \cdot C$.
- For U3 the output is $A \cdot B \cdot C \cdot D$.

In each case the output is true, or 1 if all inputs are true, or 1. Alternatively, the output is false if any one or more inputs are false.

With a logic function any number of inputs is possible, but here there is a definite limitation which depends upon the type of logic technology used. Four inputs is a comfortable number, although some devices allow up to eight.

The OR gate

The OR function is represented by the OR gate. In Figure 3.3, as before, the inputs are applied to the left-hand side and the outputs are taken from the right.

- For U4 the output at Y is $A + B$.
- For U5 the output is $A + B + C$.
- For U6 the output is $A + B + C + D$.

LOGIC GATE SYMBOLS

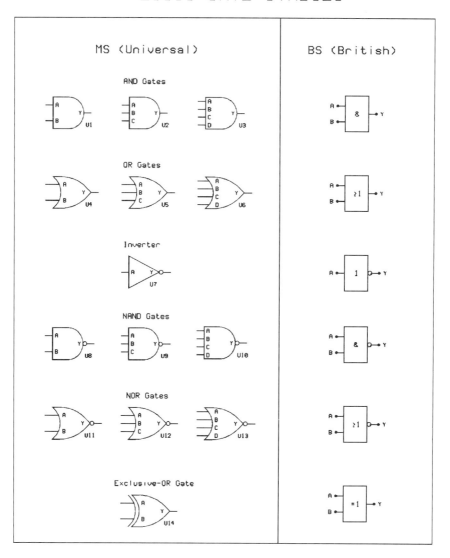

Figure 3.1

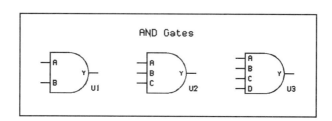

Figure 3.2

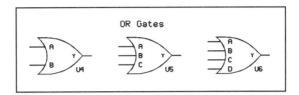

Figure 3.3

Here the output is false, or 0, if all inputs are false, or 0. Alternatively, the output is true if any one or more inputs are true.

As with AND gates, four inputs is a comfortable number, but some devices allow up to a maximum of eight.

The NOT gate, or inverter

The complement, NOT or inversion of a function is represented by the inverter (Figure 3.4). This is the only meaningful logic function that has a single input. In this case $\overline{A}$ appears at output Y.

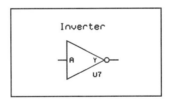

Figure 3.4

The circuit symbol is based on that for an electronic buffer or amplifier circuit. The circle on the output represents the inversion or complement and indicates that electronic inversion occurs at the output stage. The inverting circles may also be found on the inputs to gates, which, as you might expect, indicate that inversion occurs at the input stage.

The AND, OR and NOT are the basic elements for all digital electronic systems. Three other gate circuits have been developed from the basic three: the NAND, NOR and Exclusive-OR gates.

The NAND gate

The NOT AND function is represented by the NAND gate, which is one of two universal gates. Referring to Figure 3.5:

- For U8 the output at Y is $\overline{A \cdot B}$.
- For U9 the output is $\overline{A \cdot B \cdot C}$.
- For U10 the output is $\overline{A \cdot B \cdot C \cdot D}$.

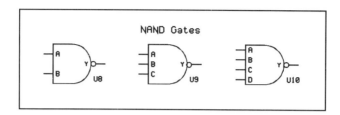

Figure 3.5

For NAND the output is false, or 0, if all inputs are true or 1. Alternatively, the output is true if any one or more inputs are false.

The NOR gate

The NOT OR function is represented by the NOR gate, which is the second universal gate. Referring to Figure 3.6.

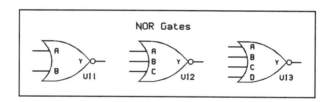

Figure 3.6

- For U11 the output at Y is $\overline{A+B}$.
- For U12 the output is $\overline{A+B+C}$.
- For U13 the output is $\overline{A+B+C+D}$

For NOR the output is true, or 1 if all inputs are false or 0. Alternatively, the output is false if any one or more inputs are true.

NAND and NOR gates are known as *universal gates* since it is possible to construct any logic function using just these two gates. The clue is De Morgan's theorem.

The Exclusive-OR gate

The Exclusive-OR, XOR or EXOR for short, is a compound function that has a special significance in digital circuitry, which we will investigate later. The function can be written as

$$Q = A \cdot \overline{B} + \overline{A} \cdot B$$

It has an output of 1 if an odd number of its inputs are 1s as shown in the following truth-table:

A	B	$A \cdot /B$	$/A \cdot B$	Q
0	0	0	0	0
0	1	0	1	1
1	0	1	0	1
1	1	0	0	0

Drawing the circuit diagram using AND, OR and inverter gates we obtain the arrangement shown in Figure 3.7. The circuit diagram for the EXOR function can be derived in a variety of different ways. Equally as valid, and perhaps more convenient, are versions comprising solely NAND or NOR gates. We will build versions of these shortly.

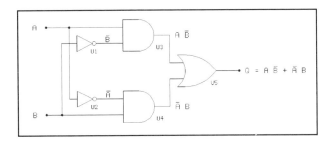

Figure 3.7

The EXOR function occurs so frequently in digital design that a special gate has been devised. The MS symbol is shown in Figure 3.8.

Figure 3.8

The Boolean function for A EXOR B is written $A \oplus B$, where the circled cross '$\oplus$' is the EXOR operator.

Points to note

- It is called Exclusive-OR because it excludes the two OR states: $\overline{A + B}$ and $\overline{A} + \overline{B}$, i.e. $A \neq B$.

Continued on p. 42

Points to note (*Continued*)

- $A + B$ is the Inclusive-OR function.
- The EXOR function only has two inputs.
- The inverse EXOR, or Exclusive-NOR, gate does exist. It generates what is known as the equivalence function and will produce a high whenever both inputs are equal. The symbol is shown in Figure 3.8 as U15.

The real value of the EXOR gate can be found by examining its truth-table. Suppose we added A and B together instead of performing the logic operation: the output column would be the same. We therefore have the makings of a digital circuit which will perform binary addition. However, we know that $1 + 1 = 2$ or, in binary, 10_2, so our new circuit will need a component to accommodate a carry. The circuit is also used in encoder and selector circuits. By just using one of the two inputs as an ENABLE or control input it is possible to output the complement of the second input variable. We will study this technique later on.

Combinational logic circuits

A combinational logic circuit is where the output of a gate is a function of the present inputs to that gate. Unlike sequential logic, these circuits are incapable of storing information.

Figure 3.6 is an example of a combinational logic circuit. To work out its Boolean function just follow these few rules:

- Make sure that each gate is numbered or can be uniquely identified.
- Starting at the input of the top left-hand gate, write the name of each signal on the input terminal.
- Write the Boolean expression for the output of each gate on its output pin.
- Work your way across the diagram from left to right, writing in each output expression. If these Boolean functions start to get complicated then just assign another (unused) letter and convert later.
- The Boolean expression for the circuit should appear on the output of the last gate in the chain.

Universal gates

As mentioned earlier these are the NAND and NOR gates. Either gate can be used to construct an AND, an OR and a NOT gate and because they are relatively easy to build form the basis of most integrated circuits (ICs). In practice a digital circuit may be built by just using one type only and because most logic ICs comprise up to four logic elements using just the one type makes economic sense.

Example

Convert the following Boolean expression to realize a logic circuit using entirely (a) NAND gates and (b) NOR gates:

$$Q = \overline{A} \cdot \overline{B} + \overline{B} \cdot D + \overline{A} \cdot C \cdot D + A \cdot B \cdot \overline{C} \cdot \overline{D}$$

Solution

First double invert, then use De Morgan's theorem to manipulate the equation (by changing the ORs to ANDs):

$$Q = \overline{\overline{\overline{A} \cdot \overline{B} + \overline{B} \cdot D + \overline{A} \cdot C \cdot D + A \cdot B \cdot \overline{C} \cdot \overline{D}}}$$

$$Q = \overline{(\overline{\overline{A} \cdot \overline{B}}) (\overline{\overline{B} \cdot D}) (\overline{\overline{A} \cdot C \cdot D}) (\overline{A \cdot B \cdot \overline{C} \cdot \overline{D}})}$$

which is a four-input NAND supplied by four NAND gates.
To convert to NOR, invert each AND element twice:

$$Q = \overline{\overline{\overline{\overline{A} \cdot \overline{B}}}} + \overline{\overline{\overline{\overline{B} \cdot D}}} + \overline{\overline{\overline{A} \cdot C \cdot D}} + \overline{\overline{A \cdot B \cdot \overline{C} \cdot \overline{D}}}$$

$$Q = \overline{\overline{\overline{\overline{A} + \overline{\overline{B}}}}} + \overline{\overline{\overline{B} + \overline{D}}} + \overline{\overline{\overline{A} + \overline{C} + \overline{D}}} + \overline{\overline{\overline{A} + \overline{B} + \overline{\overline{C}} + \overline{\overline{D}}}} \ldots$$

De Morgan gives for NOR functions.
Removing the double inversions we get a four input OR function:

$$Q = \overline{A + B} + \overline{\overline{B} + \overline{D}} + \overline{A + \overline{C} + \overline{D}} + \overline{\overline{A} + \overline{B} + C + D}$$

By inverting the whole expression twice we get our NOR circuit:

$$Q = \overline{\overline{\overline{A + B} + \overline{\overline{B} + \overline{D}} + \overline{A + \overline{C} + \overline{D}} + \overline{\overline{A} + \overline{B} + C + D}}}$$

Practical exercise: logic gates

Now we come to the interesting part. This section will introduce you to the basics of EASY-PC Professional XM (henceforth referred to as EASY-PC Pro) and PULSAR, so at the end of this chapter you should be able to use your computer to draw a simple logic circuit and simulate it.

Make sure that both EASY-PC Pro and PULSAR have been installed correctly and that your directory paths include both these applications. Consult Chapter 9 for details of how to do this if it has not been done already. It is also a good idea to operate from a WORK subdirectory so as to keep your files separate from those that are supplied.

The mouse should be in control of the cursor in both EASY-PC Pro and PULSAR. Remember that the left-hand button acts as the ENTER or RETURN key, referred to as ENTER in this text and the right-hand button is the ESCAPE key, referred to as

Continued on p. 44

Practical exercise: logic gates (*Continued*)

ESC . Commands are issued either by accessing the menu areas at the top of the screen or more directly by using keyboard commands. You must refer to the key-strip for a complete listing of these. This text will use fast keys wherever possible. For combinational key strokes the text will use SHIFT to indicate the SHIFT key, CTRL the CONTROL key and ALT the ALTERNATE key.

Creating a schematic circuit

From the DOS prompt call up EASY-PC Pro by typing EPCPROX /L-. You are greeted with the Mode Selection menu (Figure 3.9). This menu can be accessed from EASY-PC Pro drawing mode by pressing SHIFT F10 or ALT X . This is important because it provides us with an elegant termination—you exit EASY-PC Pro by typing X at this menu.

```
           Mode Selection
    Schematic              A
    PCB                    B
    Sch Component          C
    PCB Component          D
    --------------------------
    Sch Symbol             E
    PCB Symbol             F
    --------------------------
    Exit To DOS            X
```

Figure 3.9

We choose the Schematic option by typing A . You should now see the whole of the 32-inch drawing area in the centre of the screen—but, before you start creating diagrams, it is a good idea to set up some initial working conditions.

Position the cursor in the approximate centre of the drawing area and zoom in to full size by keying 5 . The dots represent 1-inch grid marks which help you to align the elements of your drawing in an organized and attractive way. By typing the command M G you now 'snap' all drawing items to a smaller 0.1-inch grid. (The smaller grid is displayed when you zoom in closer.)

Adding a component

Let us now select a gate element from one of the libraries. The sequence is very straightforward, just follow these few steps:

Step 1: F8 is the 'new component' key (Figure 3.10).

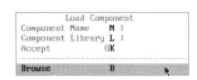

Figure 3.10

Figure 3.11

Step 2: Browse will bring up the list of current component libraries (Figure 3.11).

Step 3: Click on PULSAR.IDX with the mouse. (If you are using the supplied disk, only DEMOLIB.IDX will be available. Use this instead.)

PULSAR.IDX is the library which contains most of the primitive logic circuit elements you will need as basic building blocks. They are recognized by the logic circuit simulator, also called PULSAR. Just pause for a moment and inspect the range of devices that are offered. You ought to recognize quite a few of them (Figures 3.12 and 3.13).

Figure 3.12

Step 4: Select 2AND from the list by highlighting it with the mouse, then clicking on it with the left mouse button.

After some disk access, a SCH(ematic) Placement Reference menu appears. This allows you to assign a component reference number and, for advanced users, add a variety of parameters.

Continued on p. 46

Practical exercise: logic gates *(Continued)*

Figure 3.13

Step 5:   If the details are satisfactory, confirm by clicking on OK or by typing K.

The menu disappears, and a two-input AND gate should appear on your screen. You can move it around the drawing area by simply clicking the new spot with the left-hand mouse button. The symbol will be fixed in position when you press ESC or click the right-hand mouse button.

Pressing 3 increases the size of the image to 'zoom level' 3 (Zm3). Take a good look at the symbol. Its circuit reference number is shown as U1 (Figure 3.14). Each connection point or terminal is marked by an 'X'. You can even display the names of each connection point by turning them 'ON' in the settings preferences:

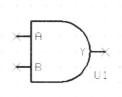

Figure 3.14

Step 6:   SHIFT F2 takes you into the menu and SHIFT N turns the pin names 'ON'.

Step 7:   ESC takes you back to your drawing.

Drawing lines

In order to simulate the gate in PULSAR you will need to give each pin a Net name. You do this by connecting short lines to each pin:

Step 8:   Place the cursor on terminal-A and press F2 for line draw. A beep confirms that connection has been made.

Step 9: Move the cursor away from terminal A and you will see a white line appear. (If it doesn't, don't worry, just press ⌧ and it will appear.) This line will rubber band as the cursor moves around the screen.

Step 10: Anchor this free end one grid point to the left of terminal A by clicking the left-hand mouse button. The line's colour changes from white to yellow.

This line has a net number which is displayed at the right-hand end of the status bar at the bottom of the screen. To simulate the gate this input net must have a name rather than a number:

Step 11: SHIFT N gives you the Net Name panel at the top of the screen, (Figure 3.15). To assign a name just type A followed by ENTER. Note the new net name on the status bar.

Figure 3.15

Step 12: ESC completes the line draw. (Figure 3.16).

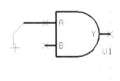

45fix 2m3 1 Net=A Width=0(10)

Figure 3.16

Repeat steps 8 to 12 for terminal B (name the net 'B') and the output, terminal Y (name it Q_AND–Don't use a space – use an underscore).

Step 13: Finally, save your work by typing SHIFT F5. As well as actually saving your work, this procedure will also generate the net-list that you will need for the next phase.

Step 14: The Save File panel offers you the current name, probably newcirc.sch together with the current directory path. Highlight the name, press ENTER and then use the left arrow key to move the cursor and modify the filename to ex1. Repeat with the path, modifying it to c:\epcprox\work (Figure 3.17). If you get a message 'Not a valid directory name', it will most probably be because the work subdirectory has not been created.

Continued on p. 48

Practical exercise: logic gates *(Continued)*

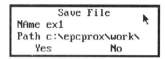

Figure 3.17

Logic circuit simulation

The next stage is to simulate this gate using PULSAR.

Step 15:   Select Logic from the Simulator section of the Tools menu. This takes you from EASY-PC Pro to PULSAR. It takes a few moments while your work is saved as a temporary file. You should get the display shown in Figure 3.18.

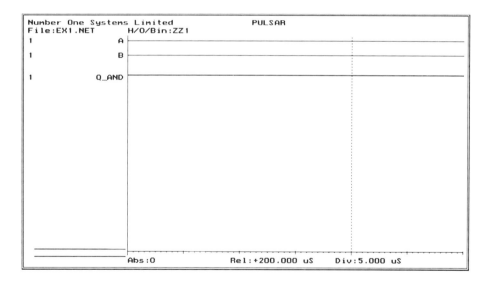

Figure 3.18

   If you get a message 'File not found' it is quite likely that your paths have not been set up correctly. Refer to Chapter 9 to correct the problem.

The PULSAR screen

We will just concern ourselves with aspects that relate to our current circuit. Note that all currently active items on the screen are highlighted in bright white. Fast-key commands are indicated by red initials on those commands.

The PULSAR screen displays two green horizontal lines which represent the two logic signal inputs *A* and *B*. The third line labelled Q_AND displays the gate's output logic.

The two vertical lines are variable cursors. The solid line is the absolute cursor and the dotted line is the relative cursor. Either is active at any one time – the one that is currently active will have its label (Abs or Rel) highlighted in the bottom menu bar.

Just above this menu bar is a time scale. Each mark on the scale has a time displacement indicated by 'Div'. A useful setting here should be 5.000 µs.

Simulation

Step 16: Click on the label A. It should change from red to white and now awaits a signal to activate the input. The input signal can either be one of a range of provided generators or an assigned square waveform.

Step 17: To apply a generator press ⎢SPACE⎥; instead of entering a generator name merely type 5 kHz. Label A turns yellow and a white square-wave trace appears in place of the green line. Note that the multiplier is case sensitive, MHz gives megahertz, mHz gives millihertz.

Step 18: Apply a 10 kHz signal to input B in the same way.

Step 19: Double click on input B to open up a space under it on the display.

Step 20: Type ⎢Z⎥ (zoom) or ⎢U⎥ (un-zoom) until you get a display like the one shown in (Figure 3.19).

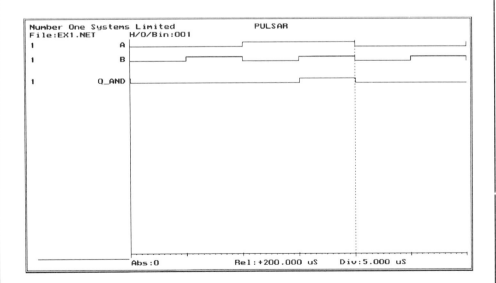

Figure 3.19

Continued on p. 50

Practical exercise: logic gates (*Continued*)

What is this screen telling us? By moving either cursor we can check each of our four logic states and map out the truth-table for AND. (If the left-hand side of your screen shows 4   2   1, click on the ⊡ to the right of File:EX1.NET, to get the column of three 1s shown in Figure 3.19.)

Step 21: ⊡ toggles between the Absolute and Relative cursors. Select Abs.

Step 22: Fix this cursor at the start by typing Ⓐ followed by Ⓞ. Absolute time is now set to zero seconds

The second menu line (Figure 3.20) displays the logic level detected by the absolute cursor either in binary (B), hexadecimal (H) or octal (O). (*Note*: If the display shows 00X, select Configuration, Analyser Settings, then Initial States. 'Undefined' will probably be highlighted. Select All HIGH, then press ⟦ESC⟧ three times to leave the menus. Answer Ⓨes to the 'Save the configuration' message.)

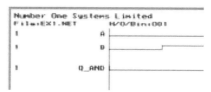

Figure 3.20

Step 23: Select the binary display by clicking on the B of B/H/Oct:. The display becomes H/O/Bin:001 and the left-hand column, which indicates the weighting of each trace, should change to 1s.

Step 24: Click on the left-hand 0 of the 001 group to give yourself a space after the colon, and on the 1 to separate that from the 0s.
The two 0s indicate the first truth-table state and the 1 the output, Q.

Step 25: Check that 'Snap' (to waveform edge) in the bottom menu bar is not active.

Step 26: Pick up the absolute cursor by clicking the mouse anywhere along that vertical line. It should now change to yellow, indicating that it has become active.

Step 27: Move the mouse pointer about 25 μs along the trace and click the left-hand button. The screen will appear as shown in Figure 3.21. The binary pattern at the top now makes more sense, but if you inspect the start of the Q_AND trace you will see that the 1 occurred just before the high/low transition. We will discuss this later.

Step 28: Move the absolute cursor along the trace in roughly 50 μs steps and you should be able to plot each state of the truth-table by recording the binary pattern.

Step 29: Return to EASY-PC Pro by typing 'Ⓠ, then Ⓨ'.

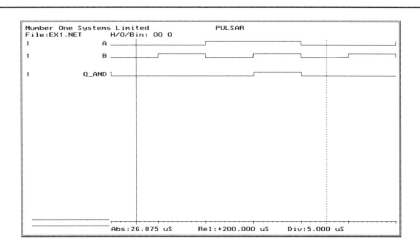

```
Number One Systems Limited              PULSAR
File:EX1.NET       H/0/Bin: 00  0
1                  A
1                  B
1             Q_AND
                        Abs:26.875 uS      Rel:+200.000 uS    Div:5.000 uS
```

Figure 3.21

Completion of the exercise

Once back in EASY-PC Pro add five other two-input gates: 2OR, 2NAND, 2NOR, 2XOR and 2XNOR. Keep the same input net names as those used on the 2AND – namely A and B. Just confirm that you are joining nets when the program asks you to do so. Some of the inputs may appear dotted after joining nets. This is a feature to show which nets are affected. When all the nets have been put in your drawing, select Unhighlight from the View menu to return them to solid lines. The output nets ought to be named separately. Use Q_OR, Q_NAND, Q_NOR, Q_XOR and Q_XNOR, respectively. The circuit diagram could appear as shown in Figure 3.22.

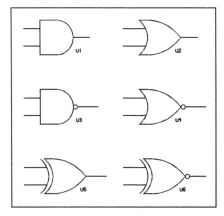

Figure 3.22

Continued on p. 52

Practical exercise: logic gates *(Continued)*

Finally, save your completed circuit, ⬛SHIFT⬛ ⬛F5⬛ ⬛ENTER⬛, simulate in PULSAR (Tools, Logic) and investigate each gate's truth-table. A likely PULSAR screen could be as shown in Figure 3.23.

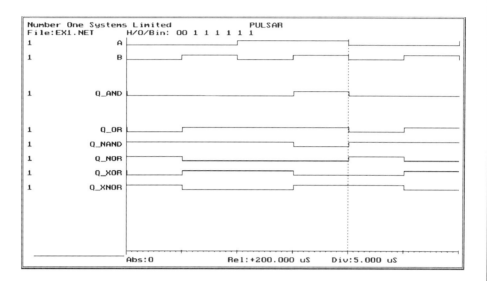

Figure 3.23

Simulation problems

If you get a message that says that your inputs are not named, then it is most likely that you have not named one of the nets properly on your schematic. Wait for the simulation to end and you may well get a clue as to which node(s) has escaped by examining the PULSAR traces. Then continue as follows:

Step 30: Quit PULSAR by typing ⬛Q⬛ ⬛Y⬛ and return to EASY-PC Pro.

Step 31: Locate the suspect input and edit the line by placing the cursor on it and key ⬛F1⬛, which is the 'edit line' key. The line should change from yellow to white. Check its net name by inspecting the bottom menu bar (Figure 3.24).

```
EASY-PC-PXM I SL Abs  15.200,  16.100 IN Grid 45fix Zm3 l Net=A Width=0(10)
```

Figure 3.24

Step 32: Change it if necessary by repeating step 11.

Step 33: If the net name is correct then it is likely that the actual connection to the gate is your problem. Place the cursor in the middle of the gate and type `F7`, the 'component edit' key. Move the cursor away roughly two grid points and click again to move the gate. If the lines are dragged along also then connections are OK – just return the gate to its original position and press `ESC` to complete the edit.

Step 34: If the lines are detached from the gate, then repeat step 8.

Step 35: To remove any unwanted lines repeat step 31 and key `D` until the line disappears.

Step 36: Save the circuit and re-simulate.

To 'quit' EASY-PC Professional XM and PULSAR

This was explained early on in the exercise.

Step 37: `Q` `Y` exits PULSAR and returns you to EASY-PC Pro.

Step 38: `SHIFT` `F10` `Y` or `ALT` `X` `Y` returns you to the Mode Selection menu after inviting you to save your current SCH file.

Step 39: At the Mode Selection menu, just key `X` to quit to DOS.

Exercises

1. Derive a Boolean expression for the output of the following circuit and draw the truth-table. Could the circuit be simplified?

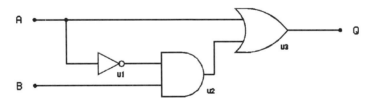

Figure Ex3.1

2. Using EASY-PC Pro, draw logic circuit diagrams which represent the following Boolean expressions:

   (i) $Q = A \cdot \overline{B} + C$

   (ii) $Q = (A + \overline{B}) \cdot (\overline{C} + D)$

Continued on p. 54

Exercises (*Continued*)

3. Using NAND gates only produce (a) a NOT gate, (b) an AND gate, (c) an OR gate and (d) a NOR gate.
4. Using NOR gates only produce (a) a NOT gate, (b) an AND gate, (c) an OR gate and (d) a NAND gate.
5. Justify the correctness of your circuits in Exercises 3 and 4 by simulating them in PULSAR to check your truth-tables.
6. In the section describing universal gates we took the following expression as an example:

$$Q = \overline{A} \cdot \overline{B} + \overline{B} \cdot D + \overline{A} \cdot C \cdot D + A \cdot B \cdot \overline{C} \cdot \overline{D}$$

Use EASY-PC Pro to draw a circuit of the function and also circuits in its NAND and NOR forms.

Part 2

4 Combinational logic circuits

Introduction

In this chapter we will examine more closely how logic gates can be connected or combined together to provide useful circuits. These combinational logic circuits are constructed with basic logic gates. Their outputs can be expressed as a Boolean equation and will correspond only to the circuit's current inputs – any previous input configuration will have no effect.

To help you tackle the design of such circuits there are a few basic steps worth following:

- Define the project to be solved.
- Draw a truth-table that conforms to the project definition.
- Derive a Boolean equation from the truth-table.
- Simplify the Boolean equation if necessary.
- Draw a schematic diagram (use EASY-PC Pro).
- Simulate the circuit (use PULSAR).

Once the logic circuit has been verified it can be turned into a circuit module and added to a library of similar functions. These will be used to assemble much larger circuit elements, typically integrated circuits.

We will consider arithmetic circuits as they embody the same techniques as decoder circuits (the other main branch of this logic).

Arithmetic logic circuits

Digital circuits are frequently required to perform arithmetic operations – addition, subtraction, multiplication and division. Since multiplication is merely repeated addition, division repeated subtraction and subtraction a case of adding a 2s complement subtrahend, the value of a binary adder circuit is obvious.

Half-adders

If we add together two binary variables, A and B, the truth-table that conforms to the operation would be as follows:

A	B	Sum	Carry
0	0	0	0
0	1	1	0
1	0	1	0
1	1	0	1

We obtain a sum twice: when $A = 1$ AND $B = 0$ and when $A = 0$ AND $B = 1$. The carry occurs when $A = 1$ AND $B = 1$ (i.e. $1_2 + 1_2 = 10_2$). The Boolean expression then becomes:

$$\text{Sum} = A \cdot \overline{B} + \overline{A} \cdot B \text{ and Carry} = A \cdot B$$

To build the logic circuit we need two inverters, two AND gates and one OR gate for the sum and one AND gate for the carry, (Figure 4.1). Alternatively, we can represent the sum part of the circuit with an Exclusive-OR gate and attach a carry section as before (Figure 4.2). The function may be illustrated as a block and can be defined as a circuit primitive, as shown on the right, in Figure 4.2. Before we do though, it may be more convenient to construct the circuit entirely of universal gates, as shown in Figure 4.3. Since NAND (and NOR) gates are available as four independent units in an integrated circuit, we would only need two integrated circuits to construct a NAND version.

In the Practical Exercise at the end of this chapter we will draw each of these circuits using EASY-PC Pro and then simulate them together using PULSAR to verify the truth-table.

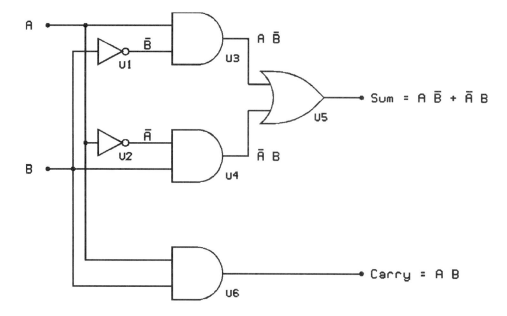

Figure 4.1

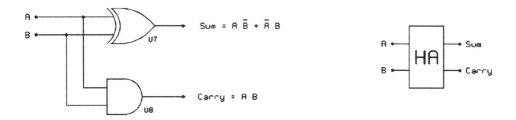

Figure 4.2

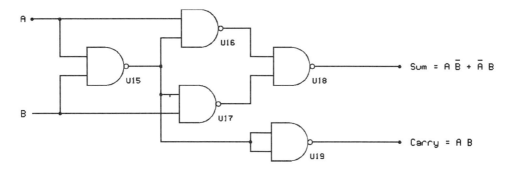

Figure 4.3

Full adder

The previous circuits may well add any pair of bits together. However, computers work with collections of bits, so a sensible adder circuit may well have to work with groups of two or more. The half-adder works fine for the least significant bit only, but we need to account for the carry when we add the next pair and successive pairs of bits. The carry now becomes a carry-out from the first pair of bits and is carried forwards to the next pair as a carry-in. A full adder therefore has as its inputs an augend, A and an addend, B as well as a carry-in C_i. Like the half-adder it will have both sum and carry-out for outputs.

To design a full adder circuit we must first draw a truth-table to represent the logic:

C_i	A	B	Sum	C_{out}
0	0	0	0	0
0	0	1	1	0
0	1	0	1	0
0	1	1	0	1
1	0	0	1	0
1	0	1	0	1
1	1	0	0	1
1	1	1	1	1

From this truth-table we obtain the following Boolean expressions:

$$\text{Sum} = \overline{C}_i \cdot \overline{A} \cdot B + \overline{C}_i \cdot A \cdot \overline{B} + C_i \cdot \overline{A} \cdot \overline{B} + C_i \cdot A \cdot B$$

$$\text{Carry out} = \overline{C}_i \cdot A \cdot B + C_i \cdot \overline{A} \cdot B + C_i \cdot A \cdot \overline{B} + C_i \cdot A \cdot B$$

These must be simplified before a logic circuit can be built. A Karnaugh map will reveal that the sum is a three-function Exclusive-OR, and the carry-out is a simple combinational logic. Proving this by Boolean algebra:

$$\text{Sum} = C_i(\overline{A} \cdot \overline{B} + A \cdot B) + C_i(A \cdot \overline{B} + \overline{A} \cdot B) = C_i(\overline{A \oplus B}) + \overline{C}_i(A \oplus B)$$

$$\text{Carry out} = C_i(A \cdot \overline{B} + \overline{A} \cdot B) + A \cdot B(C_i + \overline{C}_i) = C_i(A \oplus B) + A \cdot B$$

If we substitute P for $A \oplus B$, the equations for the full adder become:

$$\text{Sum} = C_i \cdot \overline{P} + C_i \cdot P = C_i \oplus P$$

$$\text{Carry} = C_i \cdot P + A \cdot B$$

which is merely two half-adders and an OR gate, (Figure 4.4).

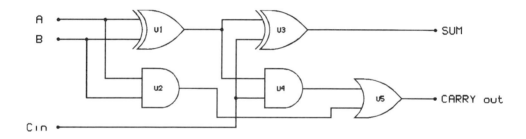

Figure 4.4

As with the half-adder it may be more convenient to build the circuit using universal gates. The resultant NAND circuit requires only nine gates and a total of 18 gate inputs (Figure 4.5) which is three fewer gates and 13 fewer gate inputs than the first implementation.

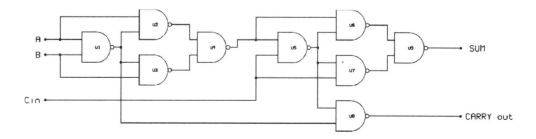

Figure 4.5

Parallel full adder circuit

Having completed the design of half-adders and full adders, the next step is to examine a practical arithmetic circuit. Using a commercially available 4-bit adder, the two binary values A1, A2, A3, A4 and B1, B2, B3, B4 may be added together. A suitable circuit diagram would be like the one shown in Figure 4.6.

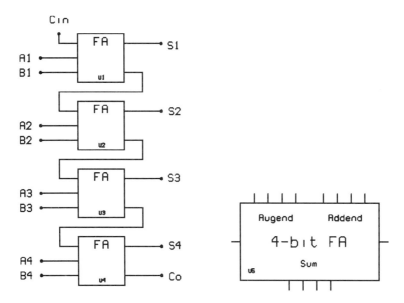

Figure 4.6

The addend would be applied to the A inputs and the augend to the B inputs. We expect to view the 4-bit result on the S outputs. This is all very fine, but we have a problem: in any electronic circuit the output is unable to change at the same instant that the inputs are applied. There is always a hold-up while the electronics stabilize. This effect is known as *propagation delay*, and we will be encountering it more and more throughout this book. In this example, the carry-out must 'ripple' through from the least significant to the most significant pair of bits and, so we have to make use of the circuit's propagation delay before there is a final carry-out. As the circuit is increased in size the propagation delay becomes far more significant.

Practical Exercise: combinational logic circuits

This series of exercises will take you through the processes of creating a schematic circuit diagram from which you will obtain a circuit net-list and eventually create a logic circuit module or primitive. You will be building upon techniques met in the Practical Exercise in Chapter 3, so be prepared to refer back from time to time.

Continued on p. 62

Practical Exercise: combinational logic circuits *(Continued)*

Creating a schematic circuit

We will start by drawing the schematic diagram of the full adder circuit, given in this chapter. The first thing to do is to start the program and initialize your screen:

Step 1:   From the DOS prompt call up EASY-PC Pro and choose the Schematic option.

Step 2:   Zoom out the screen to Zm5 by typing 5 .

Step 3:   Snap to half grid by typing M H .

Step 4:   Set track angle to 90° by typing SHIFT , A , b .

Now you need to draw a circuit that looks something like the one shown in Figure 4.7:

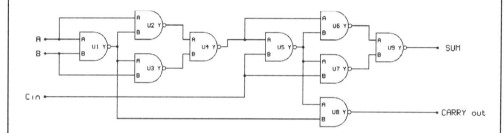

Figure 4.7

Adding components

First of all place nine 2NAND gates in roughly the same order as shown in Figure 4.7.

Step 5:   F8 is the new component key – select a 2NAND from the PULSAR.IDX library (or the DEMOLIB.IDX library if you are using the trial version supplied with this book).

Step 6:   Move the cursor to some convenient grid point on your diagram then press ENTER to position or reposition the 2NAND gate.

Step 7:   To place subsequent 2NAND gates, move the cursor to a new grid position and press R for 'repeat'. A new 2NAND should appear without you having to access the PULSAR library. Repeat the process until nine such gates are in place. Press ESC to fix the last gate in position.

Step 8:   If you are not satisfied with any of the positions, a single gate can be 'picked up' again by first moving the cursor near its input A and then pressing F7 . That component now becomes mobile and can be moved elsewhere on the screen or even deleted. The full range of operations can be found by dropping down the 'Edit' menu (Figure 4.8).

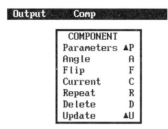

Figure 4.8

Adding tracks

The next stage is to link the gates together with 'tracks'. But before we do it is worth saying a bit more about EASY-PC Pro tracks. EASY-PC Pro tracks have a start point, called 'home' and an end point called 'end'. In the line edit mode, H takes the cursor 'home' and E to the 'end'. Both can be reversed by typing SHIFT I. Each bend or junction is at a 'node' point and you can move the cursor along the track, from node to node by typing N (for 'next'), whereas B (for 'back') moves the cursor backwards towards home.

Step 9: Starting from U1, place the cursor on terminal A and press F2 for 'line draw'. A beep confirms that connection has been made.

Step 10: Move the cursor towards terminal A of U2 and you will see a white line appear. There is bound to be a bend in the track and this should be a right-angle if you have done step 4 correctly. If the bend is in the wrong direction, F (FLIP) should reverse this.

Step 11: Anchor the free end on terminal A of U2 by clicking the left-hand mouse button or pressing ENTER. The line's colour changes from white to yellow.

Step 12: Continue the process until all connections are made.

Should mistakes occur, and they're bound to, F1 gets you into the 'line edit' mode. If you pull down the centre menu, this time you will find a range of editing options (Figure 4.9).

Step 13: Finish off the drawing by naming the three input nets A, B and C_i and the two output nets S and C_o. (*Hint:* Use F1, then SHIFT N.) You could also use Tools, Junction Dots if you want visible 'blobs'.

Continued on p. 64

Practical Exercise: combinational logic circuits *(Continued)*

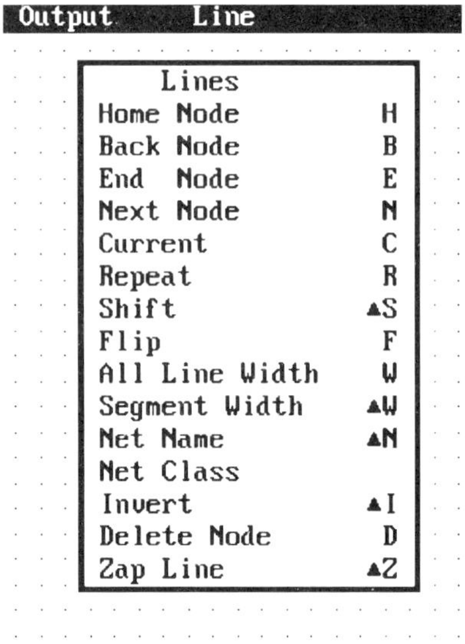

```
┌─Output────Line──────────────┐

        ┌──────────────────────┐
        │      Lines           │
        │  Home Node        H  │
        │  Back Node        B  │
        │  End  Node        E  │
        │  Next Node        N  │
        │  Current          C  │
        │  Repeat           R  │
        │  Shift           ▲S  │
        │  Flip             F  │
        │  All Line Width   W  │
        │  Segment Width   ▲W  │
        │  Net Name        ▲N  │
        │  Net Class           │
        │  Invert          ▲I  │
        │  Delete Node      D  │
        │  Zap Line        ▲Z  │
        └──────────────────────┘
```

Figure 4.9

Step 14: Save your work. Do this by typing `SHIFT` `F5` then overtyping the filename given (probably newcirc.sch) with FA_NAND. This will also create a net-list called FA_NAND.NET. Make sure that the files are saved to your work directory.

Adding text

Before you tackle the next phase of the project you may wish to embellish your diagram with some helpful information. Using the New Text facility you can put visible names on all your terminations.

Step 15: `F6` is the New Text key. Enter your text in the given panel at the top of the screen and press `ENTER` to transfer it to the drawing area.

Step 16: You will need to alter its size and width before fixing it into position—`S` `3` should be sufficient here.

Step 17: Place the text in the required position with `ENTER` and press `ESC` to complete the action.

Step 18: Continue to label each termination either by clicking the left-hand mouse key – `ENTER` or by pressing `F6` again. It will not be necessary to set size and width each time as these settings will be remembered until different parameters are required.

If you need to alter or reposition any text, ▭F5 is the Edit Text key. The Text menu gives a range of options, as shown in Figure 4.10.

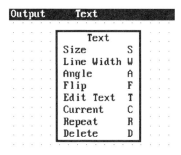

Figure 4.10

Logic circuit simulation

The next stage is to simulate the circuit in PULSAR. Selecting Logic – PULSAR from the Tools menu takes you to PULSAR. You should get a display like the one shown in Figure 4.11 (note that the order of the signals may differ). We will be referring to points on the display so you need to check back to the Practical Exercise in Chapter 3 to remind yourself of some of the information the screen is giving you. Some details are repeated here, but several more are added.

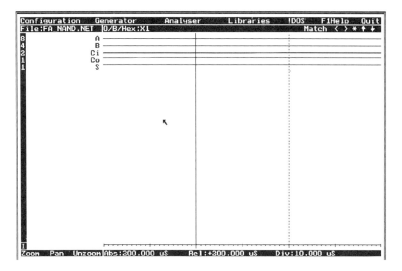

Figure 4.11

Continued on p. 66

Practical Exercise: combinational logic circuits *(Continued)*

The two vertical lines are the cursors. The one which is currently active is shown in bright white letters on the bottom menu bar. You can swap them over by typing ⌷x⌷. The bottom menu bar also has a time scale and shows how far each cursor is away from the simulation start or t_0 (Figure 4.12).

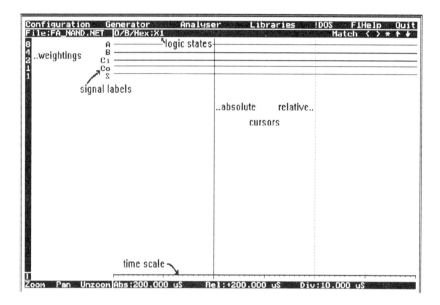

Figure 4.12

Working from the top of the screen: the H/O/B line displays the weightings of the traces that coincide with the current cursor (click on this to see how it changes its weightings). This display is useful for plotting your truth-tables, but you may not want to have all the traces included in the table, so clicking on a 1 in the vertical 'weightings' column will turn that row's weight off. (If you are not in binary mode, you may have to click on different numbers more than once to turn them off.)

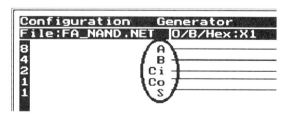

Figure 4.13

Now let us return to our simulation. At the moment the traces are not organized in any particular way (Figure 4.13) and we need to arrange them in a logical sequence.

Step 19: Click on the C_i label – it should turn white which indicates that it is now active.

Step 20: Reposition C_i by clicking on the top-most label; C_i should now be on the second row.

Step 21: Pick up A and put that trace under C_i.

Step 22: Continue until your display looks like the one shown in Figure 4.14.

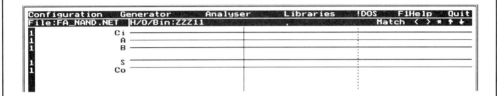

Figure 4.14

Two clicks on a label will push those below it down one row and give one row space. Alternatively, to remove a space between rows, click on that space below the label.

Now we want to apply some signals to the input traces. At the moment C_i, A and B are unconnected – this is indicated by Zs (high impedance) in the H/O/B$_{in}$: panel. (Click on B if you are in another mode.)

Step 23: Click on C_i – it should change from red to white.

Step 24: Press ⬛SPACE⬛ to call up a generator and type in 2.5 kHz The green line should change to a white square wave.

Step 25: Use ZOOM (⬛Z⬛) or UNZOOM (⬛U⬛) until one complete 2.5 kHz cycle is displayed.

Step 26: Assign 5 kHz to A and 10 kHz to B by repeating step 24. You should end up with a display like the one shown in Figure 4.15.

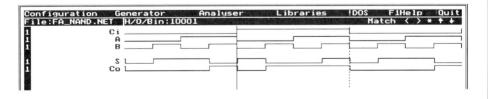

Figure 4.15

Continued on p. 68

Practical Exercise: combinational logic circuits *(Continued)*

Step 27: You can check the name of each applied generator by clicking on the chosen label and pressing $\boxed{?}$. Pressing $\boxed{?}$ again restores the signal name. Press $\boxed{\text{ESC}}$ to remove the highlight.

Step 28: Select the absolute cursor by pressing $\boxed{\text{x}}$ and send it back to the origin by entering $\boxed{\text{A}}$ followed by $\boxed{\text{0}}$. Pick up the relative cursor and position it at the end of the 2.5 kHz cycle. This should be relatively easy if Snap is turned on (use $\boxed{\text{s}}$).

Step 29: The $H/O/B_{in}$: panel now looks like 00011. (*Note*: If it does not, try clicking just to the left of the relative cursor – you have snapped to the wrong edge.)

Step 30: Click on the left-hand 0 to give yourself a space after the colon and on the 1 to separate that from the 0s.
The three 0s indicate the first truth-table state, representing C_i, A and B and the 11 the outputs C_o and S: **H/O/Bin: 000 11**.

Step 31: Now turn off Snap (to waveform edge) in the bottom menu bar with $\boxed{\text{s}}$.

Step 32: Pick up the absolute cursor by clicking the mouse anywhere along that vertical line. It should now change to yellow, indicating that it has become active.

Step 33: Move the mouse pointer a short way out to the right and click the left-hand button. You should get the trace shown in Figure 4.16.

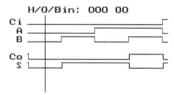

Figure 4.16

$H/O/B_{in}$: now displays the first state: i.e. $0 + 0 + 0 = 00$.

Step 34: Repeat step 33 to obtain the next state, and repeat until all states have been recorded (Figures 4.17 to 4.19).

Step 35: Verify that what you have recorded matches with what was developed in theory.

As what you have recorded should be correct, and because no curious or unexpected signals show up, you can be confident that the circuit is satisfactory.

The next stage is to translate the circuit into a Primitive module and save it as a new building block. We will then use this building block to create more complex circuits.

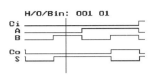

Figure 4.17

Figure 4.18

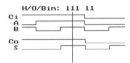

Figure 4.19

Creating a primitive component

(*Note*: This section requires a full copy of PULSAR or PULSAR Professional. The trial copy supplied with this book will not permit modification of libraries. Complete versions of the components are supplied in PULDEMO.PLB.)

Once your circuit has been successfully tested in PULSAR it can be converted into a module by adding it to a PULSAR library. Before this is done you have to make sure that no internal nodes are named – they should all be numbered. If you are uncertain about this then leave PULSAR for now and examine the file FA-NAND.NET. Only the inputs and outputs to the circuit should be named. A suitable version of FA-NAND.NET is provided at the end of this chapter. If you had to alter anything then be sure to retest the net before continuing with the exercise.

Step 36: Add the net-list to a 'PULSAR' library – click on the Libraries panel (Figure 4.20) on the top 'PULSAR' menu; you are then asked to supply a name (Figure 4.21).

Figure 4.20

Continued on p. 70

Practical Exercise: combinational logic circuits *(Continued)*

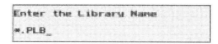

Figure 4.21

Step 37: It is unwise to add your module to one of the system libraries, so you need to create one of your own. Press HOME and enter USER.PLB (Figure 4.22).

Figure 4.22

Step 38: A USER library will be created and opened (Figure 4.23). The title bar is highlighted. Press ENTER for the library options menu (Figure 4.24).

Figure 4.23

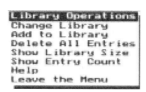

Figure 4.24

Step 39: Select Add to Library.
Step 40: You are offered the current net-list name (Figure 4.25). Accept this with ENTER.
Step 41: Now you need to give the module an entry name. Because all libraries in the path are visible to each other, the name you choose needs to be unique or else there will be a conflict. Call it FA (Figure 4.26) and then press ENTER.
Step 42: You will get a message 'Now adding to library – please wait' while it is added. Browsing the library contents will list the new addition.

```
Enter the Circuit Name:-
FA_NAND.NET_
```

Figure 4.25

```
Enter the Entry Name:-
FA_
```

Figure 4.26

Step 43: `ESC` leaves the Library utility; then `Q`uit, `Y`es takes you back to the EASY-PC Pro SCH edit screen.

Creating a schematic symbol and component

(*Note*: This section requires a full copy of EASY-PC Pro. The trial copy supplied with this book will not allow component creation. Complete versions of the components are supplied in DEMOLIB.IDX. However, symbols may be created and saved as described below.

If you want to use this new PULSAR module in future schematic diagrams you will have to create a special graphic for it. There are two parts to this. First, you have to create a schematic shape or 'symbol', and then add to this to build a schematic component. This is done in EASY-PC Pro.

Step 44: Quit Schematic edit with `SHIFT` `F10` then `Y`, and return to the Mode Selection menu.

At this point, it is a good idea to have sketched the outline of your symbol in your log-book. Roughly work out the dimensions of your outline and decide where all external connections are to go. A rectangle is a popular shape and can be oriented in either portrait or landscape (Figure 4.27). (*Note*: the scale line is included in Figure 4.27 as a guide only.)

Schematic symbol

Step 45: You should tackle the schematic symbol first. From the Mode Selection menu select `E` (Figure 4.28).
Step 46: From the Symbol Operations menu select New Symbol (Figure 4.29).

Continued on p. 72

Practical Exercise: combinational logic circuits *(Continued)*

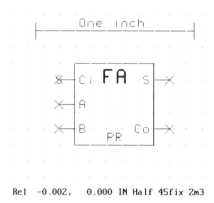

Figure 4.27

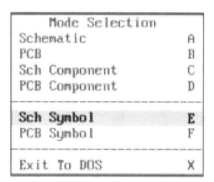

Figure 4.28

```
   Symbol Operations
   New Symbol      N
   Edit Symbol     E
   Browse          B
   Press ESC to quit
```

Figure 4.29

Step 47: The screen clears and a new menu bar appears. Select the Symbol option. Next, we must enter the symbol name. It is easier (but not at this stage essential) if this is identical to the one assigned in the PULSAR library (FA) because that node-list needs eventually to be assigned to the outline you are about to create.

Step 48: Highlight ⟨N⟩ame, press ENTER , and then type FA– ENTER in the edit box.

Step 49: A tick will appear on the left-hand side of the ⟨N⟩ame row.
This is useful as a check to ensure that at the end of this editing process all operations have been completed.

Step 50: Assign a schematic reference label by typing D and from the Reference Type menu select U, selection code B (Figure 4.30). The Schematic Symbol menu should now look as shown in Figure 4.31.

```
Reference Type
R     1   CONN  H
VR    2   LK    I
C     3   TP    J
L     4   PL    K
T     5   SK    L
D     6   J     M
ZD    7   OTL   N
LED   8   V     O
BR    9   UA    P
Q     A   UB    Q
U     B   UC    R
IC    C   UD    S
XTAL  D   UE    T
FS    E   IN    U
SW    F   OUT   V
RL    G   GND   W
```

Figure 4.30

```
           Schematic Symbol
√  ¦ Name              ¦ FA
√  ¦ Dflt Reference    ¦ U
.  ¦ Values
   ----------------------------------
.  ¦ Symbol Origin
.  ¦ Reference Origin
   ----------------------------------
.  ¦ Add to Library
   ----------------------------------
       Quit              OK
```

Figure 4.31

Step 51: You now need to draw the symbol outline, so press ESC . The menu disappears and the whole drawing area is revealed.

Step 52: Zoom to level 5: (press 5) which is roughly full size – you may proceed to draw and edit the outline in exactly the same way as you drew your circuit diagram.

Step 53: Snap to full grid: M G .

Continued on p. 74

Practical Exercise: combinational logic circuits *(Continued)*

Step 54:   Fix an origin or relative reference point – `SHIFT` `R` toggles between absolute and relative. Check this on the status line at the bottom of the screen. Select a grid point near the centre of the screen and fix the origin by typing `O`.

The co-ordinates on the status line should display: Rel 0.000, 0.000.

Step 55:   Lay down a 1/2-inch square rectangle of track, width 0 (5), checking the dimensions on the status line as you draw.

Step 56:   On Zoom 1, check that the two ends of the line you have drawn meet up (you don't want any gaps) and `ESC` to finish the rectangle.

Step 57:   Now draw the external connections to the box – you may need to go to quarter grid to do this: `M` `Q`.

You are still in the line drawing mode (check the L in the status line), so there is no need to key `F2` again – simply click on and draw these extensions. No more than 0.1-inch lengths should be needed. Once all the extensions have been drawn, connecting points or pins have to be placed on the end of each one.

It is a good idea to decide upon your origin or pin-1. This will be your starting position, and from here you can place pins in order, clockwise or anti-clockwise, around the outline. The number order is not important for schematic symbols.

Step 58:   `F4` places a pin. A collection of crosses and a number 1 appear on the screen. These show the positions of the pin number and the pin name. Make sure that the large X coincides exactly with the position of the connecting point.

Step 59:   Now edit the positions of the pin number and pin name – `SHIFT` `O` allows you to do this.

Step 60:   From the Text Positions menu select Pin Name (`SHIFT` `N`) and place the text inside the box about one grid point away from the extension. Repeat click with the right mouse button until satisfied, then `ESC`.

Step 61:   Now position the pin number by typing `N`. This time the characters 999 appear. These should be placed roughly half a grid above the pin. Again, repeat click with the right mouse button until satisfied, then `ESC` twice.

With the first pin in place you can now 'repeat' for every other pin on the outline. You will have to edit justification and positions again when you get to pin 4 (Figure 4.32).

Step 62:   In Zoom 3 you should be able to see the whole outline. Position the cursor over the next pin and type `R` to repeat.

Step 63:   If necessary, adjust the text once again. If you are careful, you may only need to adjust the text when you reach pin 4.

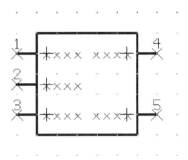

Figure 4.32

Step 64: Should you need to adjust a pin or its text once it is placed, you can return to its position and edit it by keying $\boxed{\text{F3}}$ to select it, then $\boxed{\text{SHIFT}}$ $\boxed{\text{O}}$ to enter text mode.

Step 65: Click on the Symbol menu or $\boxed{\text{SHIFT}}$ $\boxed{\text{F10}}$, then $\boxed{\text{Y}}$es to return to the Schematic Symbol menu.

To complete this phase, next you must fix the positions of the reference label and symbol origin. The symbol origin should invariably be placed on pin 1 of any module. This is important because it is the 'pick-up' point when the component is being edited − and you must know where to find it.

The circuit reference is normally placed outside the outline, well away from any connection points; but this is merely a personal preference. Figure 4.33 shows it inside; but this is just for clarity when it is used later in the book. During schematic edit, the reference label may be picked up and moved elsewhere.

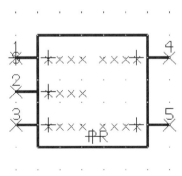

Figure 4.33

Step 66: You should be in the Schematic Symbol menu. To position the reference label select $\boxed{\text{R}}$, then click the cursor on the desired position.

Continued on p. 76

Practical Exercise: combinational logic circuits (*Continued*)

Step 67: To fix the outline's origin, select ⌑s⌑ from the menu and click on the desired position.

Now we're nearly done!

Step 68: You can write a legend in the box, although this should not be necessary if EASY-PC Pro naming conventions have been observed. ⌑F6⌑ selects new text and you can type something relevant – maybe FA in this case (Figure 4.34). Make the text large and dominant, but without swamping everything else. (Try size ⌑s⌑ ⌑5⌑, width ⌑w⌑ ⌑3⌑.)

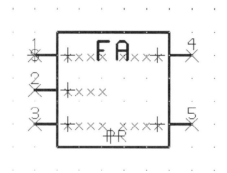

Figure 4.34

Last, but not least, save the outline by adding it to a library.

Because of the way EASY-PC Pro works internally, it is not possible to delete a symbol from a library. Save your working copies of the symbol to a working library, TEMP.SIC. When you are satisfied with your symbol, save it in the final library, for instance USR-PRIM.SIC. (Don't save it to one of the libraries supplied with EASY-PC Pro, as it could be destroyed if a new version of the program is installed.) TEMP.SIC can be deleted later.

Step 69: Select ⌑A⌑, then in the Symbol Library panel type TEMP.
Step 70: When this is complete, exit Schematic Symbol by typing ⌑Q⌑, ⌑Y⌑.
Step 71: ⌑ESC⌑ from the Symbol Operations panel and you should be back in the Mode Selection menu.

Schematic component

Step 72: Having created the symbol, you now need to create a schematic component. Choose option ⌑c⌑ in the Mode Selection menu (Figure 4.35).

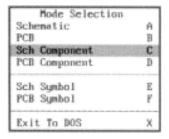

Figure 4.35

Step 73:   From the Component Operations menu, select ⃞N⃞ew: (Figure 4.36). The screen clears and a new menu bar appears. Select Component and you should get a Schematic Component menu (Figure 4.37). All you have to do is work from the top of this menu.

```
Component Operations
New                    N
Edit                   E
Browse                 B
---------------------------
Delete                 D
---------------------------
Press ESC to quit
```

Figure 4.36

```
        Schematic Component
√  ¦ Name            ¦ FA
√  ¦ Description     ¦ Full Adder
√  ¦ Load Symbol     ¦ FA
.  ¦ No. of Gates    ¦ 1
√  ¦ Package Type    ¦ DSC
.  ¦ Values
----------------------------------------
.  ¦ Pin(s)
----------------------------------------
.  ¦ Add to Library
----------------------------------------
        Quit              OK
```

Figure 4.37

Continued on p. 78

Practical Exercise: combinational logic circuits *(Continued)*

Step 74: Assign a name – this *must* be FA to correspond to the PULSAR module.

Step 75: Device description in this case is Full Adder.

Step 76: Load the symbol you have just created as follows. $\boxed{\text{L}}$oad Symbol. $\boxed{\text{B}}$rowse. Pick USR_PRIM.SIC library and choose FA (it may well be the only one there!).

Step 77: Your symbol now appears on the screen. Accept it with $\boxed{\text{K}}$, or $\boxed{\text{ESC}}$ from the symbol panel to continue editing your new component.

Step 78: Ignore No of Gates. This is important when you are generating components that will be translated onto PCB.

Step 79: You need to assign a Package Type or it may default to something unexpected – choose DSC for discrete component.

Step 80: The Schematic Component menu returns. $\boxed{\text{K}}$ or $\boxed{\text{ESC}}$ allows you to proceed with the edit. Select Name Pin from the Pins menu (Figure 4.38), or press $\boxed{\text{N}}$ with the cursor over pin 1.

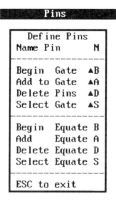

Figure 4.38

Step 81: Name pin 1 with the exact style used in the net-list (Figure 4.39).

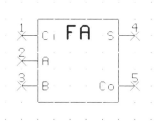

Figure 4.39

Step 82: Working around the outline, name each pin in turn. ENTER will now pick up, and ESC release.

Step 83: When you have named all the pins, re-enter the Component menu, and add the component to a schematic component library by keying A .

Step 84: The component library allows you to delete items. Nevertheless, it is wise to use TEMP.IDX as a working library until you are happy with the design. The component could then be saved to USR_PRIM.IDX Quit by typing Q uit, Y es and ESC to return to the Mode Menu.

Testing your component

Step 85: At this point it is a good idea to check that this new module works.

Step 86: Select Schematic – A and place the new component on the drawing area at Zoom 3 or 4;

Step 87: Use F8 B , select TEMP.SIC, and then FA.

Step 88: Check that the image appears and that all the pins are named correctly. (There won't be any numbers yet.)

Step 89: Check the origin or pick-up point.

Step 90: Check that the placement reference number U1 is present.

Step 91: To simulate the component you will need to connect and name dummy nets to each termination. Refer back to step 8 in the Practical Exercise in Chapter 3, and use net names the same as those appearing on the component pins.

Step 92: Save the drawing with some name such as TEST – this is not essential, but if you don't EASY-PC Pro will default the filename to NEWCIRC.

Step 93: Logic – PULSAR from the Tools menu takes you into PULSAR where you can devise a suitable test.

Step 94: Choose the same data as you used for the full circuit version and you should get the same results.

Exercises

Note: It will not be possible to complete all these examples unless full copies of EASY-PC Pro and PULSAR (or PULSAR Professional) are available. Appropriate modules are included with DEMOLIB.IDX and PULDEMO.PLB if the trial version is being used.)

1. Using your FA primitive, draw and test the following circuit for a 4-bit parallel full adder.

Continued on p. 80

Exercises (*Continued*)

2. Create a primitive module for the circuit in Exercise 1. for example:

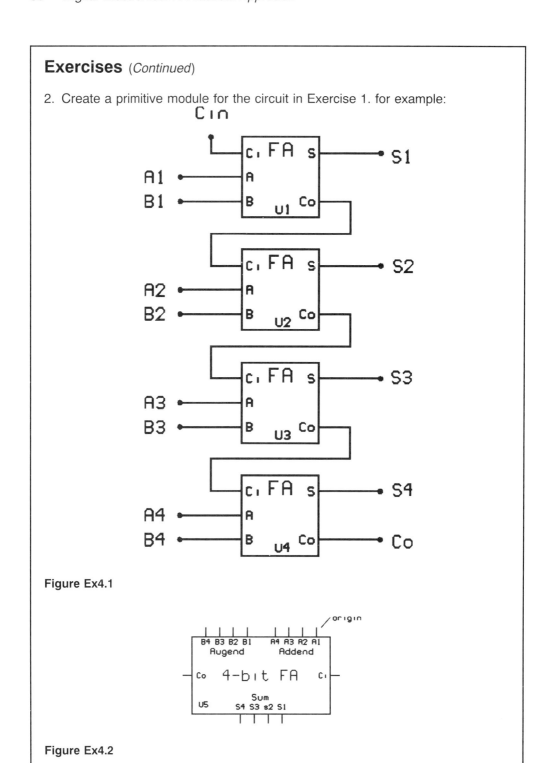

Figure Ex4.1

Figure Ex4.2

3. Draw and test the programmable 4-bit adder/subtractor circuit shown below.

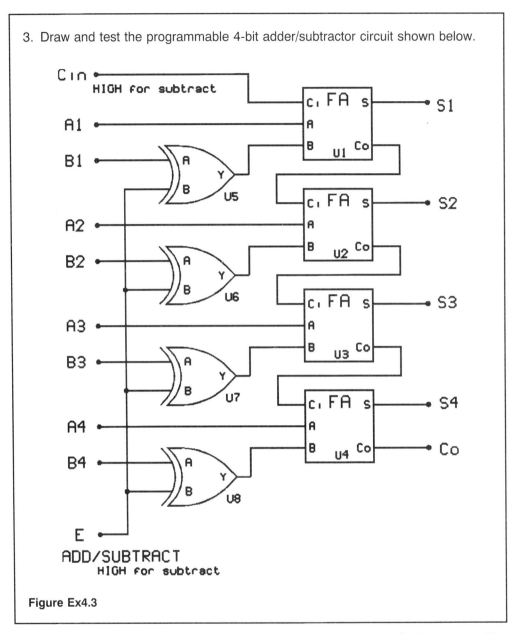

Figure Ex4.3

Continued on p. 82

Exercises (*Continued*)

4. Draw and test the 4-bit multiplier circuit shown below.

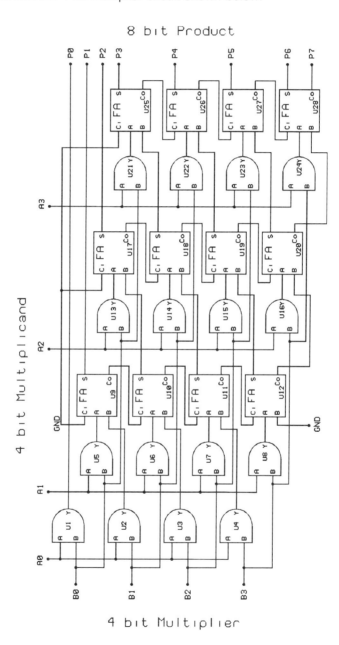

Figure Ex4.4

Net-list for full adder FA_NAND.NET:

/U1[2NAND]
[A = A]
[B = B]
[Y = 1]
_DESTPACK:
#

/U4[2NAND]
[A = 2]
[B = 3]
[Y = 4]
_ DESTPACK:
#

/U7[2NAND]
[A = 5]
[B = C_i]
[Y = 7]
_DESTPACK:
#

/U2[2NAND]
[A = A]
[B = 1]
[Y = 2]
_DESTPACK:
#

/U5[2NAND]
[A = 4]
[B = C_i]
[Y = 5]
_DESTPACK:
#

/U8[2NAND]
[A = 5]
[B = 1]
[Y = C_o]
_DESTPACK:
#

/U3[2NAND]
[A = 1]
[B = B]
[Y = 3]
_DESTPACK:
#

/U6[2NAND]
[A = 4]
[B = 5]
[Y = 6]
_DESTPACK:
#

/U9[2NAND]
[A = 6]
[B = 7]
[Y = S]
_DESTPACK:
#

Part 3

5 Sequential logic circuits: flip-flops

Introduction

So far we have only examined combinational logic circuits where outputs specified by a truth-table or a Boolean equation appear as soon as inputs are connected. The previous output states of these combinational circuits will have no effect on their current or future behaviour. On the other hand, the outputs of sequential logic circuits depend upon previous inputs as well as existing and previous outputs. In addition, the order in which the inputs are applied is of significance – hence the term 'sequential'. These circuits have the ability to retain information, and therefore form the basic building blocks for memory systems.

Bistables, latches or flip-flops

The basic building block for a memory system is known as a bistable, so-called because it has two stable states which can be changed or alternated by applying appropriate inputs. The terms latch and flip-flop are also used. Although the three terms are not identical, the differences between them are very subtle. 'Flip-flop' is the most popular designation. These then are the basic logic circuits which are required to realize the SET–RESET action, and can be constructed using universal gates.

The set–reset principle

A *toggle* switch when switched to one position may turn a lamp on and in the other position turn it off. In a set–reset system, the light is turned on by closing a SET switch. It remains on even if the switch is opened again. In other words, any further operation of SET has no effect while the lamp is on. The only way to turn the lamp off is to operate a RESET switch, which, in turn has no effect while the lamp is off, (Figure 5.1).

In a positive logic system, SET is an input S of logic level 1 and gives an output Q of logic 1. RESET is an input R of logic level 1 and gives an output Q of logic 0.

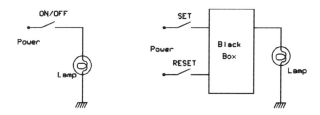

Figure 5.1

The set–reset (SR) flip-flop

This is the most simple sequential circuit element. The theoretical circuit has two inputs, SET (S) and RESET (R) and two logically opposite outputs called Q and Q'. Both NAND and NOR configurations are possible. The basic NAND circuit is shown in Figure 5.2. Here the output equations are

$$Q = \overline{\overline{S} \cdot Q'} \text{ and } Q' = \overline{\overline{R} \cdot Q}$$

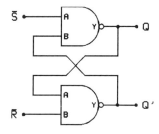

Figure 5.2

by applying De Morgan's theorem these convert into

$$Q = S + \overline{Q} \text{ and } Q' = R + \overline{Q}$$

for the NOR.

Because there are two inputs there can only be four output states. These are:

SET	RESET	OUTPUT	Operation
0	0	NQ'	The rest or stable state
0	1	0	Resets Q to 0
1	0	1	Sets Q to 1
1	1	?	Illegal because Q and Q' must be different.

If $S = R = 1$ were to be forced, then the output state would have no meaning as $Q = Q'(= 1$ for NAND, 0 for NOR). The circuit would become stable only when one of the inputs returned to 0. If both fell simultaneously, the output state would be indeterminate.

Practical SR flip-flops

The two versions of the SR latch are shown together with the logic symbol in Figure 5.3. Inverters are normally added to the outputs of the NOR version and to the inputs of the NAND version.

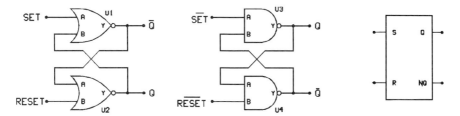

Figure 5.3

Operation

Let us examine the NOR circuit more closely.

- With SET off, a signal (logic 1) is applied to the RESET input on U2 and output Q is seen to go low. Remember that the NOR logic prescribes that if any input sees a logic 1 then the output of that gate will be logic 0.
- Input B to U1 follows the output of U2, which is also logic 0. The circuit is now latched in the RESET state, which will be sustained even if RESET were to be switched off and on repeatedly.
- With RESET off, a signal (logic 1) is applied to SET input and NQ (previously Q') is seen to go low. Input A to U2 is now also LOW. With both inputs to U2 now LOW, the output U2 is forced high – Q is therefore exhibiting a logic 1. The circuit is now latched into a SET state, which will be sustained even if SET were to be switched off and on repeatedly.
- If both SET and RESET inputs were applied together, the outputs of both gates would try to go low. This contradicts the definition that NQ is the complement of Q. So this state, SET = RESET = 1 is not allowed.

If we examine the timing diagram using PULSAR (Figure 5.4) we may get some clues as to what is going on. Initially, the output is in a SET state so when a SET signal

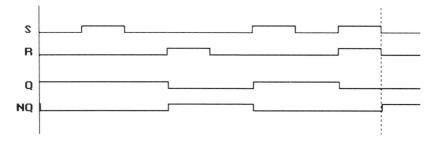

Figure 5.4

is applied there is no change on the output, Q. When RESET is applied the outputs flip and the circuit changes to a RESET state. SET is applied and once again the outputs flip. Finally, both SET and RESET are applied together. We do not expect a sensible output as that state is defined as being indeterminate. In our example, PULSAR shows both signals to be low, which contradicts the logic. (*Note*: RESET goes low fractionally before SET, so the circuit is left SET in the final condition. A few nanoseconds the other way would have given the opposite condition.)

State table

Since time is now a very important component, a revised method of representing a truth-table must be used. This new method is called a state table where:

- Q_n means a logic output level at some time t_n or prior to applying a new input condition; and
- Q_{n+1} means a logic output level during the next time interval, t_{n+1} or after applying a new input condition.

The truth-table for the SR flip-flop now becomes:

S	R	Q_{n+1}	Action
0	0	Q_n	No change
0	1	0	RESET
1	0	1	SET
1	1	?	Not allowed

Applications

Applications which require that a bistable takes some particular initial state require additional circuits to force the output to a starting value. This could be a mechanical switch (RESET button!). One useful application of the circuit is as a buffer to overcome contact bounce when a mechanical switch is operated. When operated the mechanical contacts do not make cleanly but vibrate for a period of $10–20 \mu s$ before finally settling. Because logic circuits invariably operate at very high speeds, mechanical switch bounce will be interpreted as several individual switch closures, which is particularly troublesome in keyboards and relay circuits. The set–reset flip-flop can be connected so that when a mechanical switch is operated the output will latch on or off.

A debounce circuit (using NOR) is shown in Figure 5.5, and its output waveforms are shown in Figure 5.6.

Propagation delay

There are two principal problems which the designers of bistable circuits have to minimize or overcome. These are the avoidance of the $S = R = 1$ input condition and propagation delay.

In a practical circuit there is a small time delay between applying an input signal and the appearance of the circuits response at its output. This is because a circuit always

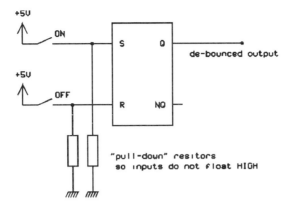

Figure 5.5

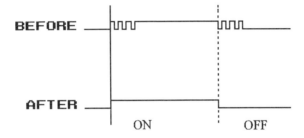

Figure 5.6

takes a finite time to operate. The phenomenon is known as propagation delay and is particularly troublesome if a chain of latches is used. We might have a situation where an output is present which has no bearing on a current input. Furthermore, this delay will give rise to hazard conditions.

Consider the simple combinational circuit shown in Figure 5.7. Each gate will have a propagation delay of about 20 ns. If we apply a logic-1 signal to input A we would expect to obtain the two outputs shown. However, if we examine the timing diagram, instead of seeing constant logic-1 and logic-0 outputs we will see spurious pulses or glitches before the output signals settle (Figure 5.8).

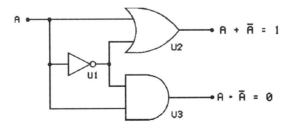

Figure 5.7

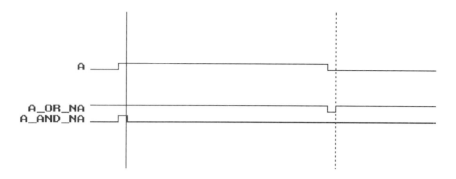

Figure 5.8

Propagation delay may, therefore, cause an output change when none should occur. In combinational circuits these unwanted signals are called *static hazards* and can be overcome by adding extra gates. These additional gates achieve the same logic but will appear redundant in a theoretical circuit.

In sequential circuits these hazards cannot be eliminated, only minimized. Now called *race hazards*, they normally occur when individual gates operate at different speeds and act together to produce unwanted input conditions. For example you might have a situation where two logic-1 signals race each other to the input of an SR bistable and both arrive together to provide $S = R = 1$! At switch-on the outputs of a latch could go to either of the Q states because one gate element would operate marginally faster than the other. Which one is which cannot be predicted, and a race condition or hazard will exist. When sequential circuits are constructed using SR flip-flops the design should be such that the condition $S = R = 1$ never arises and thus Q and NQ are always different. In addition, race conditions must be identified and excluded. As these two problems are tackled, the simple SR flip-flop will evolve into quite complicated circuits. We shall follow the development of the SR flip-flop into the more universal JK bistable.

The clocked or gated SR flip-flop

One way of tackling the problem of propagation delay in an SR flip-flop is to predict exactly when the output changes occur. We could trigger the SR action by applying pulses at some recognized time or, if several bistables are used in a circuit, their outputs could be synchronized to change simultaneously on the application of a common gating pulse.

By adding extra circuit elements, an SR bistable can be turned into a synchronous circuit (Figure 5.9). Output changes can only occur when the clock input is high; both the S and R inputs are ignored until that instant. Once *clock* $= 1$, both gates U1 and U2 are enabled and S and R can operate as before.

The timing diagram shown in Figure 5.10 illustrates this nicely. In the timing diagram, S (SET) is applied during the low part of a clock. The output Q changes immediately the clock goes high – as it would if the signal had been applied at any

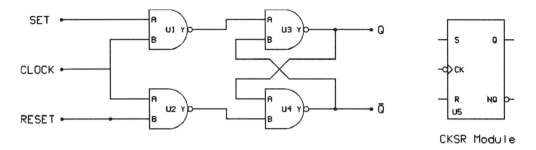

Figure 5.9

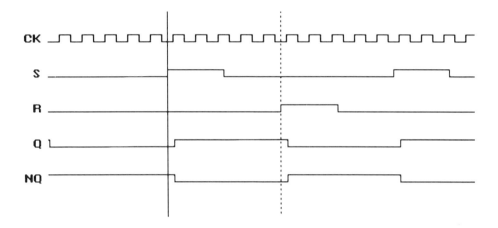

Figure 5.10

time during the high part of the cycle. The circuit is said to be *positive-level sensitive*. The clock input therefore is the gate or control which may be used to cause the circuit to operate at a predetermined time.

The operation of the circuit can be shown as a truth-table:

Clock	SET	RESET	Action
0	0	0	No change
0	1	0	No change
0	0	1	No change
0	1	1	No change
1	0	0	No change
1	1	0	SET
1	0	1	RESET
1	1	1	Not allowed

Nothing happens during the first five states on the table. These are known as the 'no change' or 'storage' states.

The state-table becomes:

Clock	SET	RESET	Q_{n+1}	Mode
0	X	X	Q_n	No change
1	0	0	Q_n	No change
1	1	0	1	SET
1	0	1	0	RESET
1	1	1	?	Not allowed

where X means 'don't care', and ? means 'not allowed' or 'don't know'.

Edge-sensitive trigger

The level-sensitive circuit is still rather limited. We have only got control over the latch during the 'on' half of a clock cycle and have not even addressed the problem of disallowed inputs. One way of getting more control over the timing is to activate the circuit only during a $0 \rightarrow 1$ (leading) or $1 \rightarrow 0$ (trailing) transition of the gating clock. These rising and falling edges of the clock represent a very small fraction of the clock period, typically 10 ns, and are the next best thing to an infinitesimally short pulse. Triggering on the edge of a clock pulse effectively excludes all spurious signals with faster duration, except those that actually coincide with the edge.

We can select the edge of our clock pulse quite easily. Consider the circuits shown in Figure 5.11. Allowing for a propagation delay of 20 ns through each inverter, those of the other gates are not significant, we obtain the pulses shown in Figure 5.12. The resulting clock now becomes a train of 20 ns pulses which are repeated every clock cycle. We merely add the desired network to the clock input of our gated SR flip-flop.

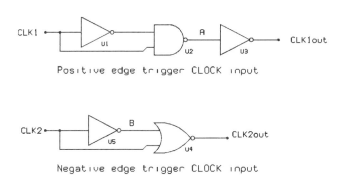

Figure 5.11

The logic symbols for gated SR flip-flops show an extra input for the clock. The edge-triggered variety has a notch-like graphic next to the clock input, and the negative-edge version has an additional inverting circle, (Figure 5.13).

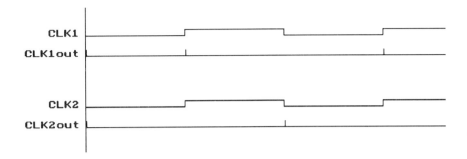

Figure 5.12

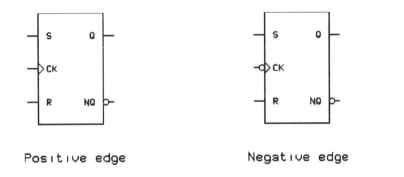

Positive edge   Negative edge

Figure 5.13

Set-up and hold times

With an edge-triggered flip-flop timing becomes very critical. Data must arrive on the input before the triggering edge of the clock pulse and it must remain there for a specified time after the triggering pulse has passed. These two characteristics are important parameters for all varieties of flip-flops.

The *set-up time* is the minimum period required for the applied signal to be constant on the input(s) before the triggering edge of the clock pulse. The *hold time* is the minimum period for maintaining a signal on the input(s) after the triggering edge of the clock pulse (Figure 5.14). The timings are usually measured at the half-way points

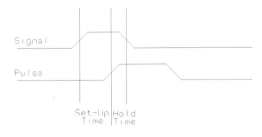

Figure 5.14

on each trace. Typical values are between 10 and 40 ns for set-up times. Hold times tend to be much shorter, and are frequently zero.

The D-type latch

One way to ensure that the condition $S = R = 1$ never occurs is to connect SET input to RESET through an inverter. This configuration is known as the delay, data or D-type latch (Figure 5.15). Because the SET and RESET are always the inverse of each other, the hazardous input condition where $S = R = 1$ cannot occur. Now we have the basis of a memory element. This flip-flop will store a single binary digit if that bit is presented at the D (data) input and a clock pulse is applied. The output Q will become the same as D until a new signal is input. The value of D is stored by the flip-flop even when it no longer exists at the input.

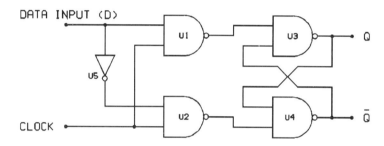

Figure 5.15

The D-type latch operation can be summarized as follows:

Clock	D	Q_n	Q_{n+1}
0	X	Q_n	Q_n
1	0	X	0
1	1	X	1

For the circuit to work effectively, the clock must be faster than the data rate expected. In the example shown in Figure 5.16, a 6 kHz clock has been used. The data input signal is a 2 kHz pulse train, offset sufficiently so that its leading edge occurs during

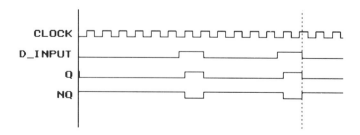

Figure 5.16

the low part of the clock pulse. Because the circuit is positive-level sensitive, the input will only take effect when the clock goes high. By enlarging the diagram we can see this more clearly (Figure 5.17). The absolute cursor shows D being applied during the low period of the clock pulse. D is transferred to the output as soon as clock goes high. The point where D goes low coincides with another high clock phase. This time the data are transferred straight away. So, for this version of the circuit, while the clock is low, the latch will be disabled and no output changes will occur.

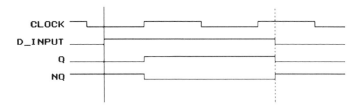

Figure 5.17

Applications

This circuit is not much use for SET and RESET operations of course, but is nonetheless a very important derivation of the SR flip-flop. There are two principal applications.

As mentioned earlier, this circuit is the basis of a memory element. One D-type latch will store one bit of data, so if a number of D-type flip-flops are connected to the same source of clock pulses then a group of binary digits is stored. A collection of eight circuit elements would be able to store 1 byte of data. This group of flip-flops is called a register and is not just available in 8s but in several sizes – typically, 16-bit, 32-bit and 64-bit as well as 8-bit.

The D-type latch may be configured as a 'divide-by-2' or 'toggle' circuit and a collection of these connected in cascade forms the basis of counter circuits. Figure 5.18 shows the configuration. Compare this with Figure 5.15. The D input has been shown as a combined S, NOT R to show the origins. Here the output from NQ is fed back to the D input and the output at Q is seen to toggle at half the rate of the applied clock. The timing diagram also monitors the output fed back from NQ as the D_{in} line (Figure 5.19).

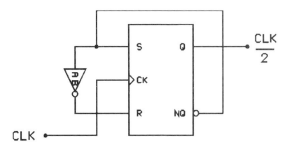

Figure 5.18

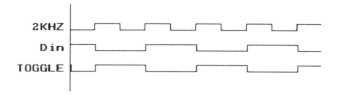

Figure 5.19

The JK flip-flop

The flip-flops we have met with so far have only tackled the problems caused by propagation delay. The JK version of the SR flip-flop addresses the indeterminate state of $S = R = 1$. This troublesome restriction can be removed by feeding back the Q and NQ outputs to the R and S inputs, respectively. The clocked SR flip-flop now has two three-input NAND gates on its input instead of the two-input types. The SET input has been renamed J and RESET input renamed K (Figure 5.20).

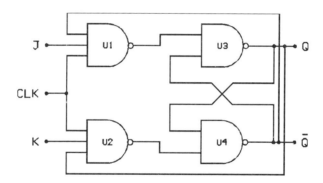

Figure 5.20

Now, when both inputs are logic 1, the output of U1 and U2 cannot simultaneously be logic 0 and the SRs input restriction is eliminated. Feedback now provides an additional toggle mode where the output of the circuit will switch at half the clock rate. This is illustrated in the PULSAR timing diagram (Figure 5.21).

Translating this into a tabular form we get:

J	K	Q_n	Q_{n+1}	MODE
0	0	X	Q_n	No change
1	0	X	1	SET
0	1	X	0	RESET
1	1	X	NQ_n	Toggle

One drawback of this circuit is that we cannot use clock pulses that are longer than the applied signal. In addition, you will find that clock width has to be limited to a

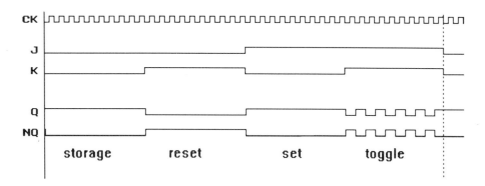

Figure 5.21

value which is dictated by the delays of the gates in the circuit. If output Q changes before the end of the clock pulse, the input conditions to J and K change again, and Q ends up being indeterminate. A way around this problem is to shorten the clock to a train of short duration pulses – by using edge-triggering circuitry.

With a pulse shaping circuit added to Figure 5.20, using PULSAR we obtain the timing diagram shown in Figure 5.22. Both J and K inputs are high, there is a 10 kHz clock from which 30 ns pulses are obtained. The outputs now change at half the clock rate.

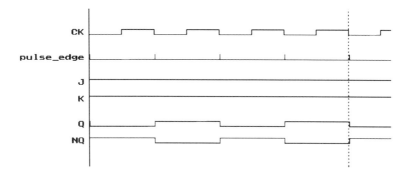

Figure 5.22

Typical timings for a JK flip-flop are:

- 20 ns set-up time,
- 5 ns hold time, and
- 40 ns propagation delay.

The master–slave flip-flop

The master–slave concept is refinement to flip-flop circuitry that eliminates the requirement for a strict clock width. This type of circuit comprises two sections (a master section and a slave section) and relies more on the turn-on and turn-off characteristics of the circuit than on the pulse width of the clock. A block diagram of an SR master–slave flip-flop is shown in Figure 5.23 – it consists of two flip-flop sections and an inverter.

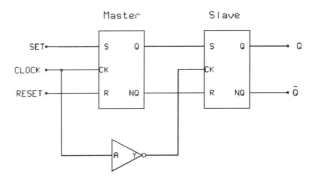

Figure 5.23

The inverter ensures that each section turns on and off alternately. On a positive clock the master flip-flop is enabled and its inputs can receive a SET or RESET. At the same time the slave section is disabled so none of the inputs can appear at the outputs Q and NQ. On the negative clock the master section is disabled and cannot sense any inputs, but the slave is enabled and transfers the outputs from the master section to Q and NQ. The timing diagram (Figure 5.24) shows the output responding during the negative half-cycle of the clock.

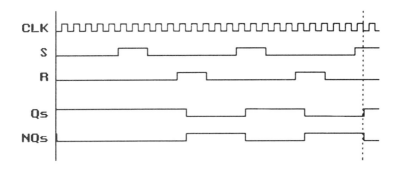

Figure 5.24

Closer inspection of the timing diagram (Figure 5.25) will illustrate what is happening. SET is applied before the clock goes positive. As soon as the clock goes high the output is latched at Q-master (Q_m), but the slave section is disabled. When the clock goes low, NCLK is high and Q_m is transferred to Q-slave (Q_s). Observe the transition delays on the outputs.

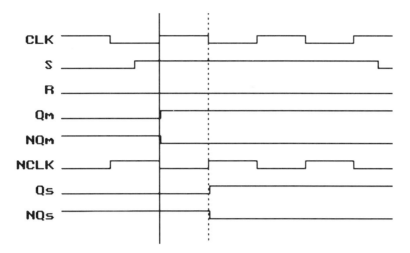

Figure 5.25

The master–slave JK flip-flop

The master–slave principle can be applied to any flip-flop version simply by introducing an inverter between the two sections. The JK version has output signals fed back from Q and NQ signals to the input gates, just as in Figure 5.20.

Figure 5.26 also shows two extra control inputs: preset and clear. The preset (NP) input allows the flip-flop to be initialized with a logic 1 (so that it starts a cycle in the SET condition) and similarly the clear (NC) input will initialize with a RESET. Adding these two control inputs only involves a minor alteration to the master section, and for NAND-based circuits will require input signals that are active low. Preset and clear

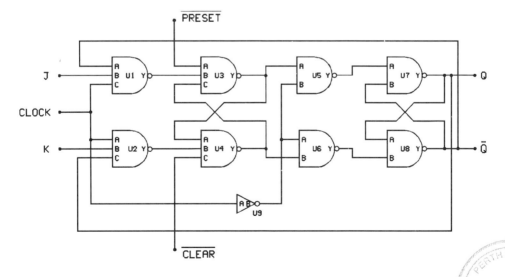

Figure 5.26

are very common control signals in digital circuits and are found on D-type flip-flop modules as well as JKs.

The circuit symbol for a negative-edge-triggered JK master–slave flip-flop is shown in Figure 5.27. The division across the middle represents the two flip-flop sections. It is quite common to find the NP and NC inputs omitted in large circuit diagrams if those inputs are not used.

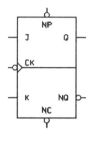

MSJK Module

Figure 5.27

Practical Exercise: flip-flops

In this section we are going to create logic circuit primitives for three of the flip-flop elements we have just considered:

• an edge-triggered JK flip-flop,
• a master–slave JK flip-flop, and
• an edge-triggered D-type flip-flop.

These three logic elements will prove invaluable over the next two chapters.

In Chapter 4 we were able to create a logic module that required little or no modification. We did this by simply drawing the schematic, simulating it and then saving the primitive in our user library. The next series of exercises will develop this approach.

Creating a primitive logic module: the JK flip-flop

First draw the schematic for the given circuit using EASY-PC Pro and then simulate your result in PULSAR. Make sure that no internal nets are named – only name the input and output terminals. (Use ⬚ then ⬚ ⬚). You can use CK (instead of CLKin), *J*, *K*, *Q* and N*Q*. To aid simulation you can name CLK_edge on both the output of the clock shaper circuit and the input to the JK circuit, but this name must be removed before you create your primitive (Figure 5.28). A suggested name for your circuit is JK_FF.SCH.

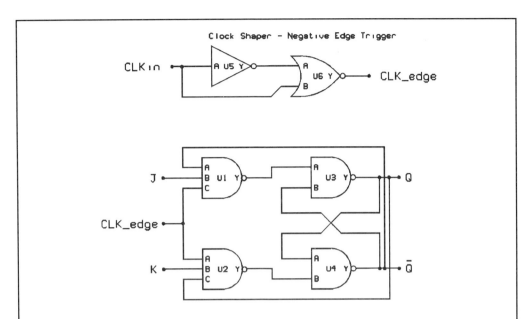

Figure 5.28

Simulation

When you are satisfied with your schematic and have saved your file, call up PULSAR from Logic – PULSAR on the Tools menu. Arrange your traces so that your screen looks something like the one shown in Figure 5.29. Apply a 20 kHz signal to CLK_in and HIGHs to J and K (Figure 5.30). It is unlikely that your circuit

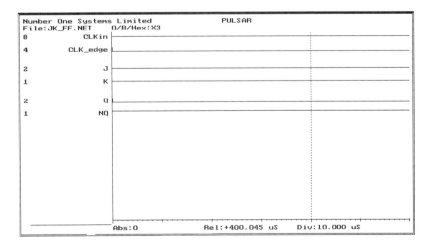

Figure 5.29

Continued on p. 104

Practical Exercise: flip-flops *(Continued)*

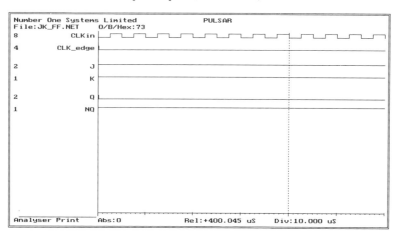

Figure 5.30

will work straightaway because only minimal timings will have been assigned to the primitives, and they will all be identical.

Introducing circuit parameters

There are three ways to introduce circuit parameters: one way is via the PULSAR editor; another is by tackling the net-list directly using a text editor, and the third is by using the value field in each component (in EASY-PC Pro). We will try the first approach initially.

Using the PULSAR editor

Step 1:   Click on File in the top menu and select Create/Modify from the Analyser Operations menu (Figure 5.31).

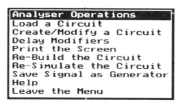

Figure 5.31

Step 2:   You will be offered the current filename, JK_FF.NET so hit ENTER to confirm.

Step 3: The next panel to appear will be a listing of all the gates used in the circuit (Figure 5.32). Highlight U5 and then press ⌷ENTER⌷ to get the Component Operations menu (Figure 5.33).

```
Circuit:    JK_FF.NET
U1          U2          U3          U4
U5          U6
```

Figure 5.32

```
Component Operations
Modify  the  Component
Delete  the  Component
Rename  the  Component
Help
Leave  the  Menu
```

Figure 5.33

Step 4: Select Modify Component to reveal current details of the inverter (Figure 5.34).

```
Component:  U5  -  INVERTER
Pin Name      Signal        Parameter     Value
A             CLKin
Y             2
```

Figure 5.34

Step 5: Click on the title bar to bring up yet another submenu, and this time select Add a Parameter (Figure 5.35).

```
Component Operations
Modify  Component  Ref.
Modify  Component  Type
Add  a  Connection
Add  a  Parameter
Show  Number  of  Pins
Show  Number  of  Parameters
Help
Leave  the  Menu
```

Figure 5.35

Step 6: Follow the instructions and enter delay followed by 30 ns on the next two panels (Figures 5.36 and 5.37). Choose 'No' parameter override for now and you should end up with the inverter panel, which now includes details of our delay parameter (Figure 5.38).

Step 7: Hit ⌷ESC⌷, ⌷ESC⌷ to return to the original PULSAR screen.

Continued on p. 106

Practical Exercise: flip-flops *(Continued)*

```
Enter the Parameter Name:-

delay_
```

Figure 5.36

```
Enter the Parameter Value:-

30ns_
```

Figure 5.37

```
Component: U5  -  INVERTER
Pin Name        Signal        Parameter        Value
A               CLKin         delay            30ns
Y               2
```

Figure 5.38

Now you have to load in the version of JK_FF.NET that you have just modified.

Step 8:   Repeat steps 1 and 2, only this time select Load a Circuit. The screen should change to the one shown in Figure 5.39. There should be positive pulses of 30 ns duration for every CLKin negative edge, but still no

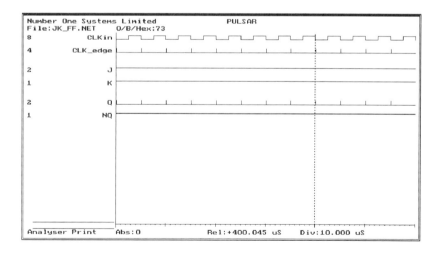

Figure 5.39

sensible activity on *Q* or N*Q* (which should be toggling). You may well have had an error message like the one in Figure 5.40, which tells you that there is more work to be done – more delays to be added to other gates in the circuit.

```
        WARNING !
  Glitch on Gate Input !
```

Figure 5.40

Using a text editor

We could add delays to every gate in the circuit by repeating steps 3 to 7. Alternatively, we could use an ASCII text editor to modify our net-list. Seeing as we need to check that no internal nodes are named before creating our library primitive, we will now tackle this second method of introducing parameters.

Quit PULSAR and EASY-PC Pro and using any text editor, load (or import) the file JK_FF.NET. (*Note*: MSDOS comes with Edit, or there is Windows' Write, Notepad or Wordpad (Win.'95). Any of these widely available editors will do; or you can use your favourite word processor, as long as you remember to save the file as 'text only' (also called ASCII format). PULSAR does not take kindly to any of the control characters that get saved in a document file!)

Once loaded, you should see a 'no frills' listing like the one below (yours will not be in three columns though):

```
/U1[3NAND]           #
[A = NQ]             /U3[2NAND]          /U5[INVERTER]
[C = CLK_edge]       [A = 1]             [A = CLKin]
[Y = 1]              [B = NQ]            [Y = 3]
[B = J]              [Y = Q]             _DESTPACK:
_DESTPACK:           _DESTPACK:          delay:30 ns
#                    #                   #

/U2[3NAND]           /U4[2NAND]          /U6[2NOR]
[A = CLK_edge]       [A = Q]             [A = 3]
[C = Q]              [B = 2]             [B = CLKin]
[Y = 2]              [Y = NQ]            [Y = CLK_edge]
[B = K]              _DESTPACK:          _DESTPACK:
_DESTPACK:           #                   #
```

Before attempting an edit, let us examine the net-list and consider some conventions:

- You will see that each circuit component is listed in its own block together with its interconnection details.

Continued on p. 108

Practical Exercise: flip-flops (*Continued*)

- Every block has its own header which comprises a circuit reference number followed by its component type (contained within square brackets).

- After this header are lines which describe the connections to each component pin. Again these are contained in square brackets. The first character is the pin name, followed by an equals sign, and then the signal name or node name to which it is connected.

- The line _DESTPACK: is generated by EASY-PC Pro. We don't need to concern ourselves with this one as it contains system information concerning the type of package used in the printed circuit board. However, this line will be followed by the component's parameter details. In U5s block you should find 'delay:30 ns', which was added earlier.

- The end of a block is marked by a # sign.

Note that lines do not contain spaces, tabs or any other control character apart from carriage return. While typing the net-list, you can mix upper and lower case letters for component references, types and names, but upper case must be used for pin names. All the alphanumeric characters on the keyboard can be used for names (which must not contain spaces) except for 17 symbols which are reserved. These are:

$$\backslash \quad | \quad / \quad " \quad ` \quad ' \quad * \quad , \quad . \quad @ \quad ^ \quad ~ \quad : \quad [\quad] \quad = \quad \#$$

Don't worry if your listing order is not identical to the one above. The differences merely reflect the order in which you actually made connections when drawing your schematic. You can see at a glance where signals have been named – there are two signal names in U6 for instance, and the delay, of course.

Now we can modify the file. But before we add delays to our gates we must carefully consider our choice of values. We know that the typical propagation delay for a JK flip-flop is about 40 ns. It seems logical therefore that each stage or gate delay should be 20 ns. However, this particular circuit relies upon its 'race' conditions to start working, so there is no point in assigning the same delay value to every gate. Varying the delays by about ±2 ns for each gate pair will avoid symmetry and ensure start-up.

To assign a propagation delay parameter to each gate you simply add the line 'delay:20 ns' after _DESTPACK: at the end of each block, remembering to vary the values by ±1 ns. You can give U6 a 20 ns delay also.

Finally, edit the node names. There is only one internal name CLK_edge and this occurs in U1, U2 and U5. Change this name to the number 4, which is the next available net number. Also, change CLKin to CK: this is to conform with a naming standard we need to use for all primitives. You should end up with something like this:

```
/U1[3NAND]              #
[A = NQ]                /U3[2NAND]              /U5[INVERTER]
[C = 4]                 [A = 1]                 [A = CK]
[Y = 1]                 [B = NQ]                [Y = 3]
[B = J]                 [Y = Q]                 _DESTPACK:
_DESTPACK:              _DESTPACK:              delay:30 ns
delay:22 ns            delay:20 ns              #
#                       #
                                                /U6[2NOR]
/U2[3NAND]              /U4[2NAND]              [A = 3]
[A = 4]                 [A = Q]                 [B = CK]
[C = Q]                 [B = 2]                 [Y = 4]
[Y = 2]                 [Y = NQ]                _DESTPACK:
[B = K]                 _DESTPACK:              delay:20 ns
_DESTPACK:              delay:18 ns             #
delay:19 ns            #
```

Save your file with the new name JKFF.NET (so as not to overwrite the original version), and return to PULSAR.

Load in the edited net-list by first keying F then SPACE to browse through available files. Highlight JKFF.NET and key ENTER to select. Because we are dealing with a new net-list the traces will be out of order again. Rearrange them so that the screen appears something like the one in Figure 5.41. Simulate the new version by applying a 20 kHz clock to CK, a 500 Hz generator to J and 1 kHz to K to obtain a screen like the one shown in Figure 5.42. Key U to zoom out until you see a complete 500 Hz cycle on J (Figure 5.43) and check that the JK flip-flop performs as described in the text. The final screen should look like the one in Figure 5.44 – the added text describes each of the JKs operating states.

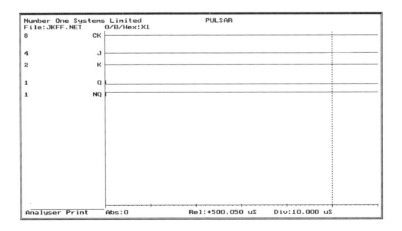

Figure 5.41

Continued on p. 110

Practical Exercise: flip-flops *(Continued)*

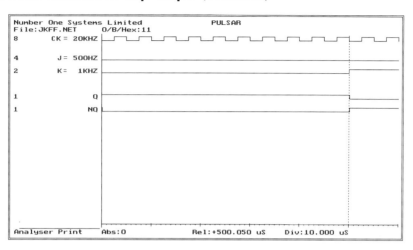

Figure 5.42

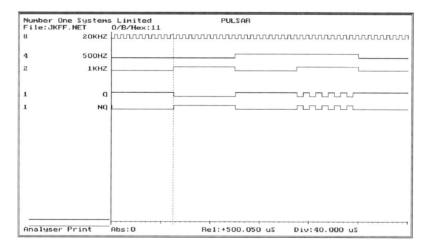

Figure 5.43

You may find that no matter how careful you are, all the traces remain red, or undefined. This is because our JK model uses internal feedback. PULSAR has a set-up option which allows all signals to be don't know, or undefined. This condition propagates through the model, feeds back, and becomes self perpetuating! The solution is to select All High from the Configuration, Analyser Settings, Initial States menu, then ESC ape back to the main screen and save the new configuration. When the simulation is satisfactory, you are ready for the next phase, which is primitive creation.

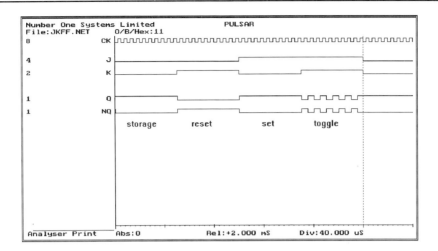

Figure 5.44

Creating a JK flip-flop primitive

(*Note*: If you are using the disk supplied with this book, you will not be able to save a new part. Please skip this section unless you have the full version of both programs.)

While still in PULSAR follow steps 4–51 in the Practical Exercise in Chapter 4. They are summarized here:

Step 44: To add the net-list to a PULSAR library click on the Libraries panel on the top PULSAR menu.

Step 45: To supply a library name, press ⎡SPACE⎤ to browse and select USER.PLB.

Step 46: When USER library opens; highlight the title bar then ⎡ENTER⎤ for library options.

Step 47: Select Add to Library.

Step 48: You are offered the current net-list name, JKFF.NET, so accept with ⎡ENTER⎤.

Step 49: Give the module the entry name JK.

Step 50: ⎡ESC⎤ leaves the Library utility.

Step 51: ⎡Q⎤uit, ⎡Y⎤es takes you back to EASY-PC Pro.

Creating a schematic symbol and component

(*Note*: If you are using the disk supplied with this book, you will not be able to save a new part. Please skip this section unless you have the full version of both programs.

Continued on p. 112

Practical Exercise: flip-flops *(Continued)*

You will need to refer to the Practical Example in Chapter 4 once more and complete the final set of steps up to step 94 therein. To assist you in this task I give you the schematic symbol and component outlines together with the final menus.

Your schematic symbol ought to be a standard flip-flop outline with a negative-edge clock input. Call it NEG_FF and use it as the basis for other flip-flops later on. (*Note*: Do not draw the scale line, as it has been included here merely as a help.)

Remember to finally save the symbol outline to the USR_PRIM.SIC library and stick to JK as the primitive name when defining the schematic component.

Your schematic symbol menu and outline will look like those in Figure 5.45 and 5.46, respectively. Drawing these little 'inversion' circles can be rather difficult. One method is to borrow the image from a suitable symbol in one of the libraries provided. The following sequence should help you.

```
                  Schematic Symbol
√   ¦ Name                ¦ NEG_FF
√   ¦ Dflt Reference      ¦ U
.   ¦ Values
-------------------------------------------
√   ¦ Symbol Origin
√   ¦ Reference Origin
-------------------------------------------
.   ¦ Add to Library
-------------------------------------------
         Quit                  OK
```

Figure 5.45

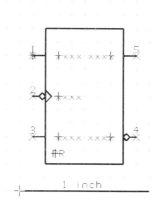

Figure 5.46

Before you start to draw the outline at step 52 (Chapter 4), you will need to import an outline from the 74LS.SIC library:

Step 9:   Zoom out to level 3: [Z] [3].
Step 10:   Snap to full grid: [M] [G].
Step 11:   Load symbol: [F8].
Step 12:   [B]rowse and select the 74LS.SIC library (Figure 5.47).
Step 13:   Choose 1OF2DSEL.
Step 14:   Snap to half grid: [M] [H].

```
                      Browse/Load Icon
         Path Name    c:\epcprox\lib\74LS.SIC
         13NANDP       17IPBITP      1OF2DSEL      1OF2DSLN
         1OF2DSP       1OF4DSEL      1OF4DSLP      1OF8DSEL
         1X2           1X21X3NV      1X2NV         1X2_1X3A
         1X3A          1X3ANV        2ANDP         2BLATCHP
         2GLATCHP      2NANDP        2NANDPSH      2NORP
         2ORP          2TO4LDEC      2X2A_2X3      2X4AOIP
         2XNORP        2XORP         2_4LDEC       3ANDP
         3NANDP        3NORP         3TO8LDEC      3TO8LDQ
         4216LDEC      4ANDORP       4ANDORP1      4ANDP
         4BITADDP      4NANDP        4NANDPSH      4REG1
         4REG2         4SHREGP       5NORP         5SHREGP
         8BITADDP      8BTSHREG      8NANDP        8SHREG1
         8SHREG2       8SHREG3       8TO3ENNP      8_3ENCKP
         8_3ENCNP      9TO4ENNP      ADDP          ALUP
         BCD2DEC       BCD7SEGP      BCD7SGNP      BCD7SGP
         BINCNT2P      BINCNT3P      BINCNTP       BINCONTP
         BITSH1P       BUFTRIP       BUFTRIP1      COMPP
         COUNT1        D12CNTP       DECOUNTP      Page Down
```

Figure 5.47

Step 15:   Use [F1] to pick up the line at connection 4. Move the mouse cursor left, then [SHIFT] [S] moves the line clear of the symbol.
Step 16:   Move the cursor to a point off the top left corner of the outline and draw a block surrounding the outline: [F10] at a corner, move to diagonal corner, then [ENTER]. Do not include the line you have just moved (Figure 5.48). The outline should now be highlighted.
Step 17:   You now need to 'zap' the contents of the block. The complete set of block options are available in the 'Group' menu (Figure 5.49). Either click on Zap or, in future, just key [SHIFT] [Z] to delete the contents of the block. The outline disappears, leaving only the graphic you want (Figure 5.50).
Step 18:   Block around this element: [F10], [ENTER], (Figure 5.51).
Step 19:   Move the cursor roughly $1\frac{1}{2}$ inches to the right and duplicate the image by keying [R] (repeat).
Step 20:   Rotate the angle through 180° by keying [A] [2].
Step 21:   You should now have two usable elements to incorporate in your schematic symbol outline. You should now be picking up the exercise from step 9 in Chapter 4.

Continued on p. 114

Practical Exercise: flip-flops *(Continued)*

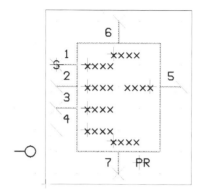

Figure 5.48

Figure 5.49

Figure 5.50

Figure 5.51

Step 22: When the outline is drawn, simply block each imported item and place in position by keying ⬚s (for shift), or use ⬚F1 and ⬚SHIFT ⬚s (for line shift).

Before saving your own schematic symbol, inspect the drawing area and check that no residual elements from the borrowed outline are present. Save it finally to USR_PRIM.SIC, then continue to create your schematic component. (Remember to use 'TEMP.SIC for intermediate saves.) Your schematic menu and component (saved to 'USR_PRIM.IDX') will look like those in Figures 5.52 and 5.53, respectively.

```
            Schematic Component
√  ¦ Name            ¦ JK
√  ¦ Description      ¦ Std JK F-F
√  ¦ Load Symbol      ¦ NEG_FF
.  ¦ No. of Gates     ¦ 1
√  ¦ Package Type     ¦ DSC
.  ¦ Values
   -------------------------------------
.  ¦ Pin(s)
   -------------------------------------
.  ¦ Add to Library
   -------------------------------------
        Quit              OK
```

Figure 5.52

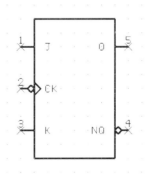

Figure 5.53

Continued on p. 116

Practical Exercise: flip-flops *(Continued)*

Creating a primitive logic module: the master–slave JK flip-flop

(*Note*: If you are using the disk supplied with this book, you will not be able to save a new part. Please skip that part of the next section unless you have the full version of both programs.)

Draw the schematic shown in Figure 5.54 using EASY-PC Pro. Use the same input and output net names as in the previous exercise. Use NP for the preset input and NC for the clear input.

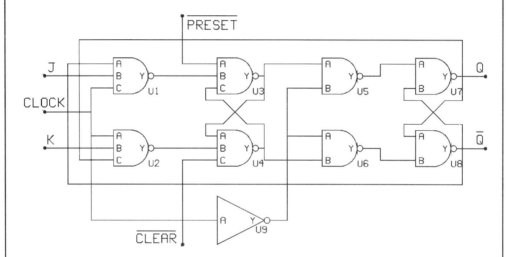

Figure 5.54

You may wish to monitor the inputs to the slave section when trouble-shooting in PULSAR. Use Qm and NQm for master outputs, and NCLK for the slave clock. These must be changed to numbers before you create a circuit primitive.

Save the circuit as JKMS_FF.SCH. Then simulate using PULSAR, employing the same signals as in the previous exercise. The timing waveform should look something like that in Figure 5.55. Note that the output changes occur on the trailing edge of the applied clock.

Make sure that the signals applied to NP and NC are high. Investigate what happens when each of these signals are taken low.

Two special generators should be supplied with your disk to test these control lines. Called clear and Preset, they both give a single 50 μs negative pulse (with different offsets) which are useful only when you use a 20 kHz clock input.

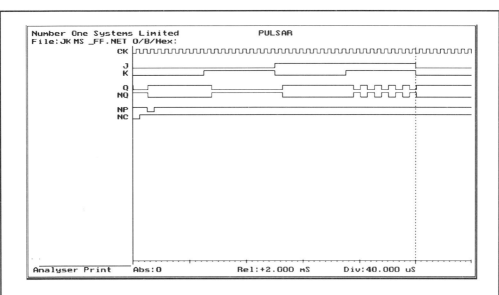

Figure 5.55

Make sure that you assign suitable delay values to each gate. The final net-list JKMS_FF.NET with no internal node names, should look something like this:

/U1[3NAND]
[A = NQ]
[C = CK]
[Y = 1]
[B = J]
_DESTPACK:
DELAY:20 ns
#

/U2[3NAND]
[A = CK]
[C = Q]
[Y = 2]
[B = K]
_DESTPACK:
DELAY:20 ns
#

/U3[3NAND]
[A = NP]
[C = 4]
[Y = 3]
[B = 1]
_DESTPACK:
DELAY:20 ns
#

/U4[3NAND]
[A = 3]
[C = NC]
[Y = 4]
[B = 2]
_DESTPACK:
DELAY:20 ns
#

/U5[2NAND]
[A = 3]
[B = 5]
[Y = 7]
_DESTPACK:
DELAY: 20 ns
#

/U6[2NAND]
[A = 5]
[B = 4]
[Y = 6]
_DESTPACK:
DELAY:20 ns
#

/U7[2NAND]
[A = 7]
[B = NQ]

[Y = Q]
_DESTPACK:
DELAY:20 ns
#

/U8[2NAND]
[A = Q]
[B = 6]
[Y = NQ]
_DESTPACK:
DELAY:20 ns
#

/U9[INVERTER]
[A = CK]
[Y = 5]
_DESTPACK:
DELAY:30 ns
#

Whilst still in PULSAR add this module to your USER.PLB library as JKMS. Quit PULSAR and create a schematic symbol and schematic component for this element. The symbol should be a standard negative-edge triggered master−slave

Continued on p. 118

Practical Exercise: flip-flops *(Continued)*

block, which you can call NEG_MSFF (saved in USR_PRIM.SIC). The component should have the same name as the library module, i.e. JKMS (saved in USR_PRIM.IDX). The two outlines should be as shown in Figures 5.56 and 5.57. Note that the component pin names are the same as those used for the PULSAR library.

Figure 5.56

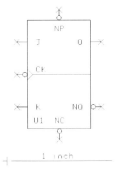

Figure 5.57

Creating a primitive logic module: the edge-triggered D-type latch

(*Note*: If you are using the disk supplied with this book, you will not be able to save a new part. Please skip this part of the next section unless you have the full version of both programs.

Investigate the schematic shown in Figure 5.58 using EASY-PC Pro and PULSAR. Call the circuit DT_FF.SCH.

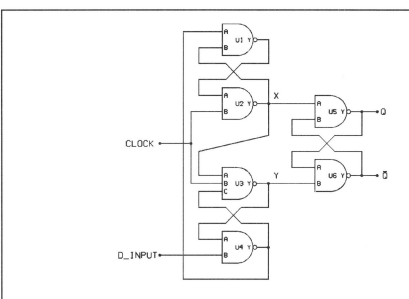

Figure 5.58

Use a 20 kHz clock and a 5 kHz data signal. Observe the gating action at points X and Y which isolates the positive clock edge.

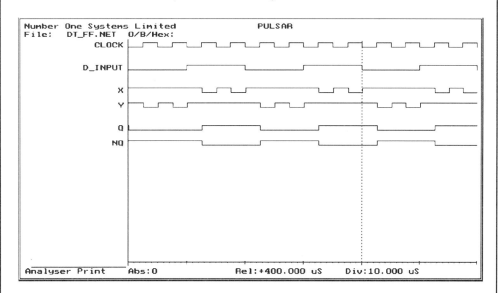

Figure 5.59

Choose a longer period data signal (1.5 kHz, say) and note that the output does not change until the first positive clock edge. Try a data signal which is more than twice the clock. Explain what happens (Figure 5.59)!

Continued on p. 120

Practical Exercise: flip-flops *(Continued)*

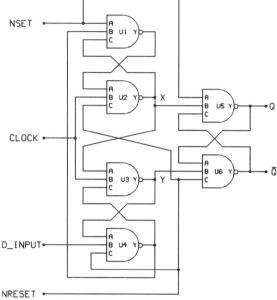

Figure 5.60

When you are satisfied with the D-types operation, redraw the whole circuit adding SET and RESET inputs as shown in Figure 5.60. Save this schematic as DT_7474.SCH.

Test the result as before, only this time use a 1 kHz data signal on D and use the Preset and Clear on generators NS and NR, respectively (Figure 5.61). The final net-list should look like this:

/U1[3NAND]	/U3[3NAND]	/U5[3NAND]
[A = NS]	[A = 2]	[A = NS]
[B = 1]	[C = 1]	[C = NQ]
[C = 2]	[Y = 4]	[Y = Q]
[Y = 3]	[B = CK]	[B = 2]
_DESTPACK:	_DESTPACK:	_DESTPACK:
delay:20 ns	delay:21 ns	delay:21 ns
#	#	#
/U2[3NAND]	/U4[3NAND]	/U6[3NAND]
[A = 3]	[A = 4]	[A = Q]
[C = NR]	[C = NR]	[C = NR]
[Y = 2]	[Y = 1]	[Y = NQ]
[B = CK]	[B = D]	[B = 4]
_DESTPACK:	_DESTPACK:	_DESTPACK:
delay:19 ns	delay:20 ns	delay:19 ns
#	#	#

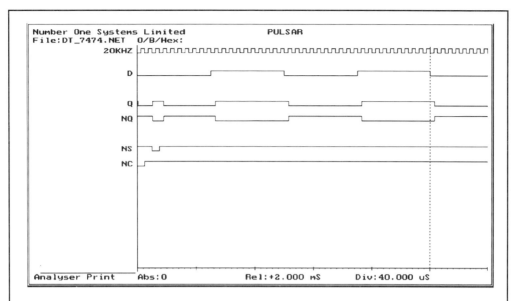

Figure 5.61

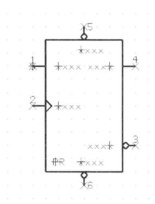

Figure 5.62

In PULSAR create a primitive module called DT7474 and save it in USER.PLB. In EASY-PC Pro create a schematic symbol called POS_DT (Figure 5.62) and save it in USR_PRIM.SIC. Create a schematic component called DT7474 (Figure 5.63) and save it in USR_PRIM.IDX.

Continued on p. 122

Practical Exercise: flip-flops (*Continued*)

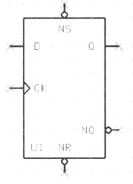

Figure 5.63

Exercises

1. Demonstrate that a D-type latch may be constructed using just four two-input NAND gates as shown below. Prove that your circuit is correct by simulating it in PULSAR (using the same settings as for the previous version).

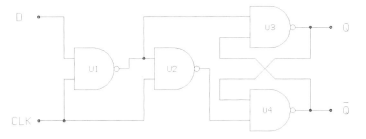

Figure Ex5.1

6 Asynchronous and synchronous counters

Introduction

In the last chapter we investigated basic sequential circuit building blocks where the order, in time, of applying inputs to these circuits was important. We established that the circuit's inputs must follow a specific sequence to produce a required output. Sequential logic systems are divided into two groups:

- A *synchronous circuit* is one where all the changes take place simultaneously at a time determined by a control signal, usually a clock.
- In an *asynchronous circuit* there is no common control. A change in one section causes further changes in other sections, and these propagate right through the circuit.

Counters are employed to keep track of event sequences. In a computer system they can keep track of the instruction sequence in a program or count the number of bits in or out of a mode converter. They can perform frequency division for delays or provide an accurate trigger for a timing circuit. Counters may count in a natural binary sequence or in some other mode. These circuits are widely available in families of MSI (medium-scale integration) logic devices and the Practical Exercises at the end of this chapter will investigate the construction of such systems.

Counter circuits are constructed by connecting toggled bistables in cascade. They also can be subdivided into two groups:

- asynchronous (or ripple) counters, and
- synchronous counters.

In a synchronous counter all the flip-flops change state synchronously with the applied clock pulse, whereas they do not in an asynchronous counter. In the latter case each flip-flop in the circuit changes state only as a result of the change in the output of another in that system.

Counters, then, change state in a predetermined pattern or sequence – called the *code*. Such codes include the natural binary, Gray code and Johnson code. The number of different states in the code is called the *modulus*.

Asynchronous counters

We will investigate asynchronous counters first because they are arguably the simpler of the two types to explain and construct. You are encouraged to build each circuit in turn, using the primitives created in the previous chapter and examine their operation using PULSAR. At the Exercise to this chapter you will be offered a variety of counter circuits to design or to investigate.

When considering counter theory, the size of the code will be restricted to just 4 bits, but the strategy can be extended to 8 bits and above quite easily.

First, we will consider the natural binary count sequence:

#	D C B A
0	0 0 0 0
1	0 0 0 1
2	0 0 1 0
3	0 0 1 1
4	0 1 0 0
5	0 1 0 1
6	0 1 1 0
7	0 1 1 1
8	1 0 0 0
9	1 0 0 1
10	1 0 1 0
11	1 0 1 1
12	1 1 0 0
13	1 1 0 1
14	1 1 1 0
15	1 1 1 1

Counting up through the 16 states, from 0 to 15, we see that in binary the least significant bit, A, toggles every state. (There is an unwritten convention that bit labelling always starts with the least significant bit.) The next least significant bit changes state when the previous bit was a 1. In other words, each more significant bit will change as its lesser significant bit changes from 1 to 0 – the count in fact 'ripples' through. Similarly, counting down is achieved as each lesser significant bit changes from 0 to 1.

We can design a circuit that can be pre-set or cleared to 0000 and then, on the application of a trigger pulse or clock, be seen to count up to 1111, re-set and started again. With simple modifications this circuit can be made to count down, or in reverse.

Asynchronous 4-bit binary counter

Four toggling master–slave JK bistables are connected in cascade (Figure 6.1). A clock signal is applied to the clock input of U1. The output of U1, now half the frequency, is connected to the clock input of U2, and so on right through the chain of flip-flops.

The count can be observed by monitoring Q1, Q2, Q3 and Q4 (the most significant bit will be on Q4). PULSAR demonstrates this nicely. Note that the NP connections

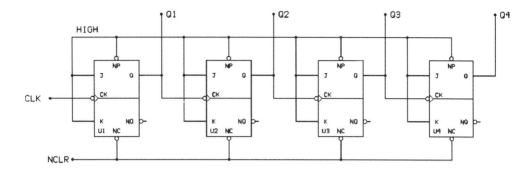

Figure 6.1

must be driven by HIGH.GEN, and NC by CLEAR.GEN. The display (Figure 6.2) shows one complete count cycle, framed by the absolute and relative cursors. The bistables' outputs change state on the trailing edge of the applied clock and data are seen to ripple through – because each flip-flop has to wait for an output pulse from the previous stage.

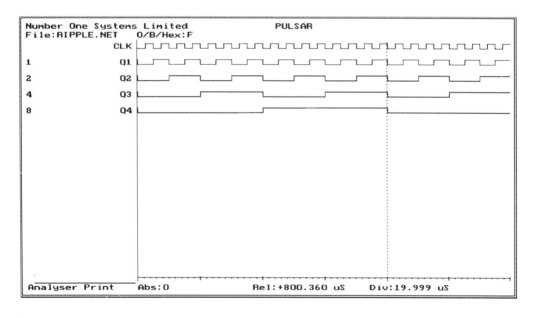

Figure 6.2

In practice this clock edge is later than the applied clock by one propagation delay period. The output at Q4 is delayed by all four previous circuits, as shown in Figure 6.3 in PULSAR. In this circuit each flip-flop has a propagation delay of 90 ns. Figure 6.3 shows the delay between the 16th clock edge and the changes in each of the flip-flop outputs, and the problem is quite apparent. PULSAR shows the overall delay between the clock edge and Q4's edge to be 360 ns (4 × 90 ns). This is the displacement of the relative from the absolute cursor.

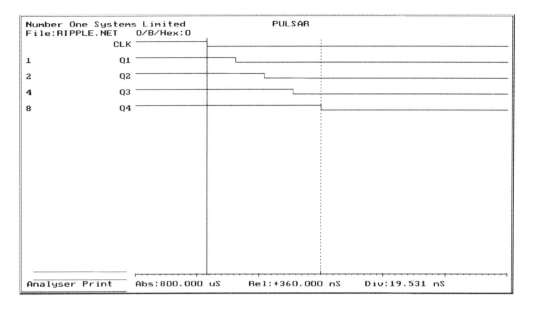

Figure 6.3

Only low operational speeds are possible due to these cumulative propagation delays (a 20 kHz clock is used here). Typically, for a 4-bit circuit with devices having delay values of 20, 25, 18 and 17 ns, the overall delay would be 80 ns and the maximum possible clock rate before the counter ceases to function correctly would be:

$$\text{Frequency} = \frac{1}{\text{time}} = \frac{1}{80}\,\text{ns} = 12.5\,\text{MHz}$$

Exceeding this rate would give rise to an unpredictable count. For our circuit, the maximum safe clock speed would be:

$$\frac{1}{360}\,\text{ns} = 2.8\,\text{MHz}$$

By replacing the 20 kHz clock with one of 6 MHz, as in Figure 6.4, we can see that the count has well and truly broken down. This is the effect of dynamic hazards, which are caused by the different transmission delays and propagation delays in the signal paths. The output signals occasionally do not comply with the predicted sequence in the truth-table, and what we get are extra counts where only one state or change is expected.

However, these circuits have the distinct advantage of being simple to implement and modify so long as the implications of propagation delay have been carefully considered at the design stage. A simple modification would be to get this circuit to count in reverse. If successive inputs were to be connected to the previous NQ outputs (Figure 6.5) then the circuit would count down (Figure 6.6). Here the circuit has been re-set at the start by applying the Clear generator to the NCLR input. Again the NP inputs are connected to HIGH.GEN.

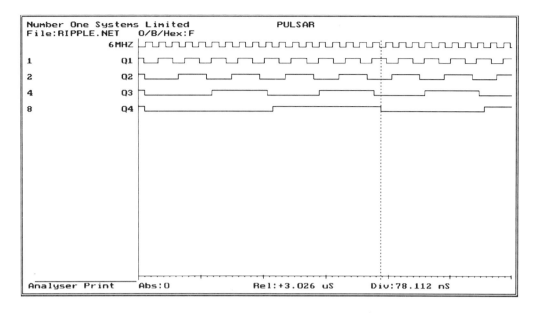

Figure 6.4

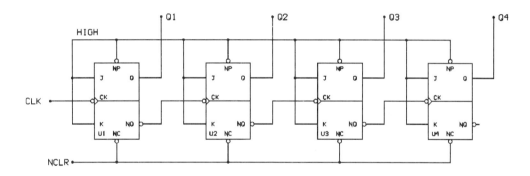

Figure 6.5

Asynchronous 4-bit binary up/down counter

Some applications require a counter to count both up and down. A typical example would be an electronic device that logs cars entering and leaving a car park. As a car enters the car park the counter would count up, and as one leaves the counter would count down, so that at any one time the total number of parked cars would be known. By adding some simple control circuitry it is possible to construct 'bi-directional' counters, i.e. counters which can count up and down.

In the circuit shown in Figure 6.7, when a logic-1 control signal is applied to UP/DOWN, U5 is enabled and U6 disabled. All the upper AND gates are enabled, which allows the Q outputs of JK to ripple through the circuit. We have an 'up' count. When a logic-0 is applied the opposite happens; all the lower AND gates are enabled and we have a 'down' count. (*Note*: If you are building this with the supplied disk,

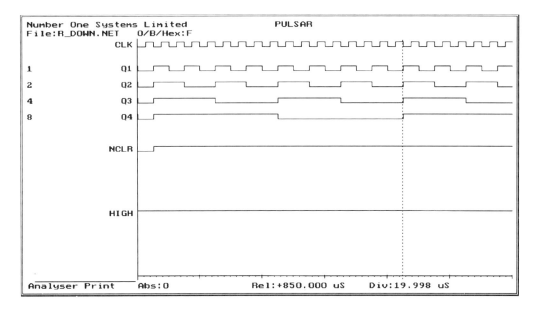

Figure 6.6

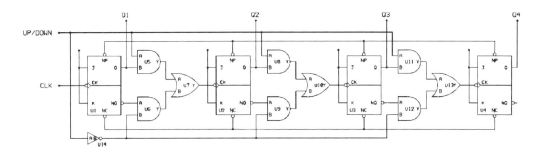

Figure 6.7

you will need to use 2X2ANDOR, a combined component, to stay within the capacity. The result will look like Figure 6.8.)

The control circuitry can be built in other ways, as shown in Figure 6.9. This can be justified by examining the logic equation for the selection logic of the circuit shown in Figure 6.7. If we call the up/down control line C, then $C = $ up and $NC = $ down. Outputs to each successive flip-flop are then $Q \cdot C + \overline{Q} \cdot \overline{C}$. So when $C = 1$, the equation becomes

$$Q \cdot 1 + NQ \cdot 0 = Q$$

and when $C = 0$, the equation becomes

$$Q \cdot 0 + NQ \cdot 1 = NQ$$

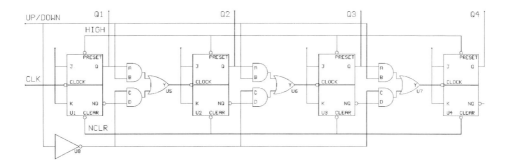

Figure 6.8

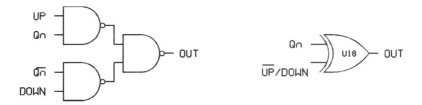

Figure 6.9

For the NAND configuration the equation becomes:

$$\overline{\overline{Q \cdot C} \cdot \overline{\overline{Q} \cdot \overline{C}}} = Q \cdot C + \overline{Q} \cdot \overline{C}.$$

The Exclusive-OR option gives the simplest configuration (Figure 6.10). Here the outputs are taken only from Q to one of the Exclusive-OR inputs. The direction is selected by either a logic-1 (down) or logic-0 (up) connected to the second Exclusive-OR input.

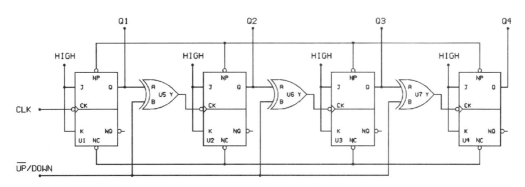

Figure 6.10

If E is the control signal, the equation becomes:

$$Q \oplus E = Q \cdot \overline{E} + \overline{Q} \cdot E$$

for $E = 1$ we get

$$Q \cdot 0 + NQ \cdot 1 = NQ$$

and for $E = 0$

$$Q \cdot 1 + NQ \cdot 0 = Q$$

Asynchronous 4-bit decade counter

In our natural binary counter all the outputs of the flip-flops are cleared to zero at the initial state. The counter sees zeros again on the 16th clock pulse and if we had an N-bit counter this zero would occur again after a count of 2^N. But we can arrange to

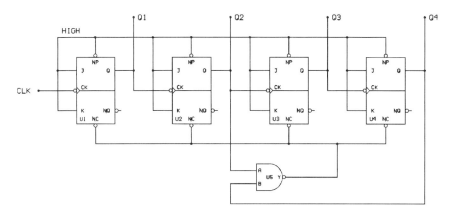

Figure 6.11

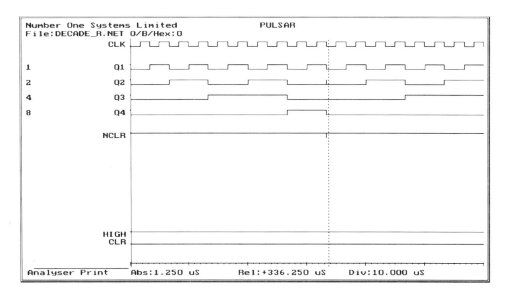

Figure 6.12

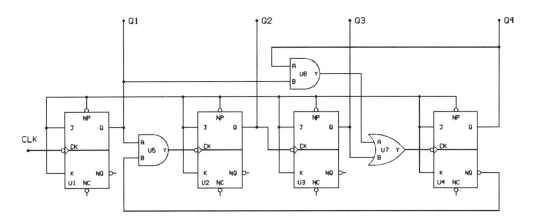

Figure 6.13

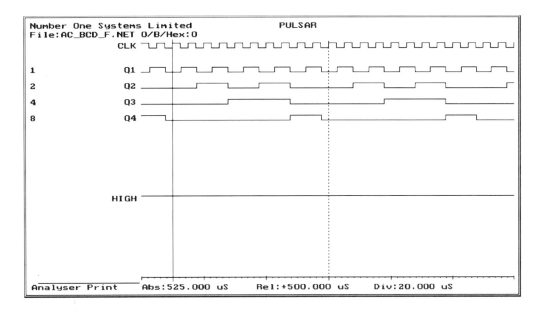

Figure 6.14

clear all the flip-flops at any other state as well, which gives rise to counter circuits of different moduli.

By adding some simple logic to the natural binary counter we can construct a decade or 8421-BCD counter (Figure 6.11). This is the re-set method. The circuit counts in the usual way, but when the 1010 output pattern is detected, the NAND gate supplies the bistables with a Clear signal, which resets the counter to 0000 after the tenth count (Figure 6.12). (Remember to connect the NP inputs HIGH.)

Alternatively, but with more thoughtful design, we can achieve a decade count without using Clear (Figure 6.13). This is the feedback method. When all outputs are zero, NQ4 is fed back to U5 which enables that gate and connects all four bistables as a natural binary counter. When 1000 is detected, U5 is disabled but U6 is enabled. Q2 and Q3 remain at zero until U5 is enabled again, two states later (Figure 6.14).

Asynchronous 4-bit modulo-N counter

By expanding the re-set concept we can design 4-bit counters that will count up (or down) to any number below 16 and stop, re-set to 0000 and count up (or down) again, simply by adding some extra logic. A modulo-12 counter would need logic to detect 1100 and provide a Clear signal; a modulo-9 counter would need to detect 1001. The feedback method requires more design effort and we shall be looking at this technique shortly.

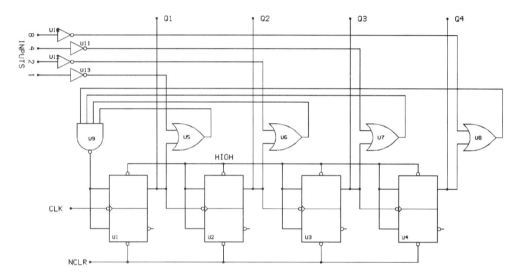

Figure 6.15

Asynchronous 4-bit variable modulo counter

It is sometimes necessary to supply a counter that can count up to any pre-set count and stop. By adding some simple external logic it is possible to input a binary pattern to the counter which when detected will stop the count. If you are building the circuit shown in Figure 6.15 using the supplied software, you will need to omit the inverters on the inputs. Remember when simulating that the inputs will be the wrong sense. In this arrangement the terminal count is input on the four inverters. When that same pattern is present on the four JK outputs the 4NAND provides a logic-0 to the input of U1; which stops it toggling and hence stops the count. (*Note*: Comparison using OR gates works because the count is ascending binary. For a reverse count, AND gates would be needed. The general case of a random repetitive count sequence would require Exclusive-OR gates to be used. It is instructive to investigate why this is the case.)

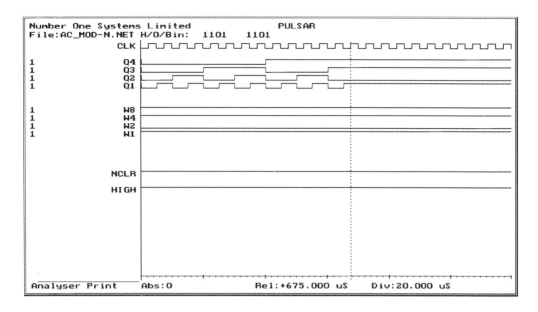

Figure 6.16

PULSAR shows this happening for a modulo-12 count, where 1101 is input on W8421 (Figure 6.16). How could this circuit be made to count continuously up to its chosen modulus, re-set and started again?

Synchronous counters

We have seen that asynchronous counters have significant problems caused by cumulative delays. These may be overcome by ensuring that when bistables change state they do so together, at the same instant in time. This is achieved by synchronizing each bistable to a common clock and using the J and K inputs to determine whether or not the bistable toggles. In this way a faster operation is achieved; the total time delay being that due to the slowest flip-flop in the circuit, plus any additional gating logic.

Synchronous 4-bit binary (up) counter

In this practical example you will need to use four of the JK master–slave flip-flops from your own primitive library plus two two-input AND gates (2AND) from the PULSAR library.

The circuit given in Figure 6.17 shows the least significant bit, Q1, on the left-hand side, but extending the circuit to five or more bits is simply a matter of connecting additional JKs and AND gates to the right-hand side. Remember to name all used inputs and outputs, including High and CLK. It is not necessary to name the Preset

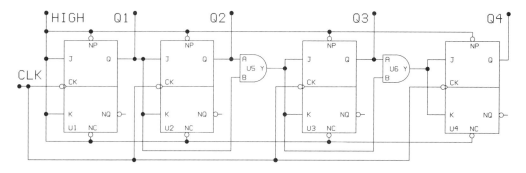

Figure 6.17

and Clear inputs this time, so just ignore the 'Inputs not named' message when loading the net-list into PULSAR. As this circuit is used as a model for future exercises you should save it as S_BIN_UP.SCH.

In order to view a sensible count sequence when simulating in PULSAR, click on the I (invert) tag in the left-hand margin (which reverses the sequence) and click on the tags for CLK and High to turn them off. You can examine the count in binary or hex by moving either cursor through each clock cycle (Figure 6.18). The bistable outputs are monitored on Q1, Q2, Q3 and Q4. Each bistable is connected to a common clock, so that state changes occur simultaneously.

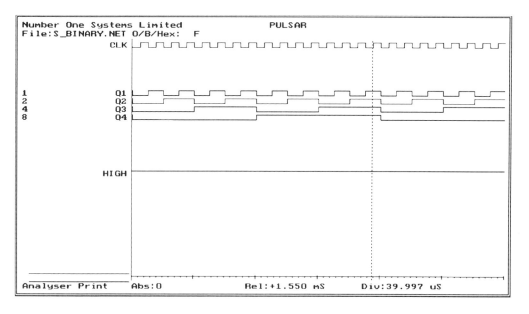

Figure 6.18

The circuit has been wired so that:

- U1 toggles every clock cycle.
- U2 toggles every 2nd clock pulse.

- U3 toggles every 4th clock pulse.
- U4 toggles every 8th clock pulse.

This means that the J and K inputs of each block must be connected thus:

Table 6.1

JK inputs	Sources
J1K1	High
J2K2	Q1
J3K3	Q1 · Q2
J4K4	Q1 · Q2 · Q3

Remember the truth-table for the JK bistable:

Table 6.2

J	K	Q_{n+1}
0	0	Q_n
0	1	0
1	0	1
1	1	NQ_n

But, more helpfully in this form

Table 6.3

Q_{n+1}	J	K
To maintain a 0	0	X
To maintain a 1	X	0
To toggle 1 to 0	X	1
To toggle 0 to 1	1	X

where X means 'don't care'.

In the circuit U1 inputs are tied high so that bistable toggles at half the clock rate and U2 toggles each time Q1 is high. The AND gates U5 and U6 ensure that U3 toggles when Q1 and Q2 are high and U4 toggles when [Q1 and Q2] and Q3 are high.

Brief inspection of Table 6.1 suggests that the size of the counter can be extended quite easily by adding extra JKs. The logic for J5,K5 would be Q1.Q2.Q3.Q4, etc. An N-bit binary counter should therefore be quite a simple circuit to build. Try extending Figure 6.1 to 8-bits.

Synchronous 4-bit binary (down) counter

We saw earlier that there are many instances when a down counter is required in a digital system. An initial state may be set up using the preset inputs and the circuit counts down towards zero as clock pulses are applied.

Like the asynchronous version, the down sequence is achieved by merely connecting JK inputs to the previous NQ outputs. This means that the J and K inputs must now be connected thus:

Table 6.4

JK inputs	Sources
J1,K1	High
J2,K2	NQ1
J3,K3	NQ1 · NQ2
J4,K4	NQ1 · NQ2 · NQ3

As before, an N-bit binary counter should be quite straightforward to build.

One refinement is suggested, however. Inspection of Tables 6.1 and 6.4 suggest that the 'sources' can be rewritten thus:

Table 6.5

JK inputs	Sources
J1,K1	High
J2,K2	NQ1
J3,K3	JK1 · NQ2
J4,K4	JK2 · NQ3

This keeps the logic simple and avoids the necessity for multiple input gates, which would be difficult with counters larger than 8-bit.

In general, the equations can be written as follows:

$$J_n K_n = J(K)_{n-2} \cdot Q_{n-1} \text{ for an UP count}$$

$$J_n K_n = J(K)_{n-2} \cdot NQ_{n-1} \text{ for a DOWN count}$$

Synchronous 4-bit binary (up/down) counter

This practical example combines the features covered in the previous two exercises. Starting with S_BIN_UP.SCH as a model, first save it as S_UPDOWN.SCH and then EDIT in the extra components.

The same UP/DOWN select logic encountered for the equivalent asynchronous circuit can be used (Figure 6.19). The design of the Exclusive-OR version is not immediately obvious by inspection, and is included as an exercise at the end of this section. The NAND version is chosen in our example because it makes use of an AND/NOT

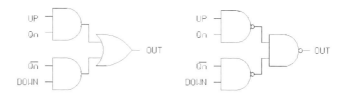

Figure 6.19

combination to realize the carry forward logic. When naming nets, call the select control E (for 'enable'). The 4-bit circuit, shown in Figure 6.20, can very easily be extended to *n*-bits if needed.

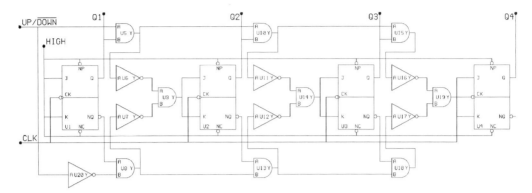

Figure 6.20

Load the net-list into PULSAR and set CLK to 20 kHz. Change the generator on E from high to low and examine the counter action.

If you are using the program supplied with this book, you will not be able to build this circuit. Instead, use the alternative form shown in Figure 6.21. The gating has been simplified, and only three stages used. It is an interesting exercise demonstrating theoretically that this behaves the same way.

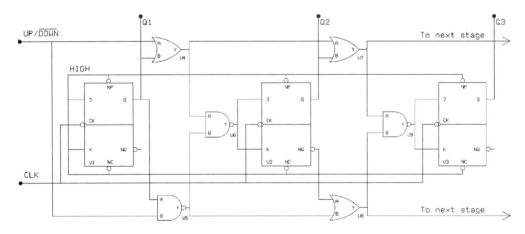

Figure 6.21

Operation

The equations for an up count and a down count can be combined thus:

$$J_1 = K_1 = 1$$
$$J_2 = K_2 = E \cdot Q_1 + NE \cdot NQ_1$$

$$J_3 = K_3 = E \cdot Q_1 \cdot Q_2 + NE \cdot NQ_1 \cdot NQ_2$$

$$J_4 = K_4 = E \cdot Q_1 \cdot Q_2 \cdot Q_3 + NE \cdot NQ_1 \cdot NQ_2 \cdot NQ_3$$

where E is the 'enable' or up/down select signal.

The select logic permits outputs to be taken from either Q or NQ. The excitation function for up is ANDed with ENABLE and that for down is ANDed with NOT ENABLE. The two corresponding composite functions are ORed to obtain the J and K equations. Therefore, when up/down select $= 1$, the equations reduce to those of an up counter, and when select $= 0$ the equations reduce to those of a down counter.

Exercise

Investigate the implementation of a circuit using the general formula:

$$J_n = K_n = JK_{n-1}(EQ_{n-1} + NENQ_{n-1})$$

where E is the 'enable' or up/down select signal, and JK_{n-1} represents either J_{n-1} or K_{n-1}. Confirm that this reduces to the expressions above.

Further refinements to these circuits could involve PRESET, CLEAR and a COUNT control signal. A COUNT signal would be ANDed with the clock and can be used to disable the clock and hold any non-zero count state. The PRESET inputs may set the counter to start at a particular state.

Synchronous 8421-BCD (up) counter with re-set

Just like the asynchronous counters investigated earlier, it is quite possible to design and build synchronous counters that work for any desired modulus. The following sections describe techniques for the construction of a BCD or decade counter, and suggests other examples.

The RESET method is the least complicated. To obtain a synchronous 8421-BCD counter therefore, Figure 6.16 could be modified by the addition of a NAND gate U7 that provides a CLEAR when the 1010 state is detected. Starting with the S_BIN_UP.SCH circuit, first save it as S_BIN_R.SCH, then use EDIT in the NAND gate U7 and its connections to NCLR. Name the high input, the four Q outputs, the clock line (as CLK) and the CLEAR line as NCLR (Figure 6.22).

Simulate the circuit using PULSAR. At first you will notice that there does not appear to be a count of ten and that there is no apparent output on Q4. By editing in a propagation delay of at least 20 ns on U7, the controlling NAND gate, and re-simulating, you should achieve the correct count.

One of the drawbacks of the RESET method as far as synchronous circuits are concerned is that the 10th state must be detected before RESET can occur. This is not much of a problem for slow counters, but becomes quite significant at high speeds. Note the negative CLR pulse on the NCLR line and the glitch on the Q2 output at the same instant.

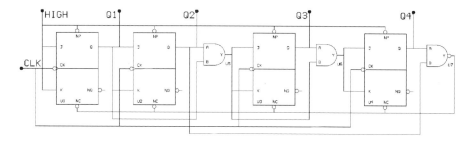

Figure 6.22

Synchronous 8421-BCD (up) counter with feedback

This is the preferred method for synchronous counter design. The method is rather more complicated than the previous one, but the design steps are quite clear. It involves drawing a state and transition table from which excitation Karnaugh maps are drawn. When minimized, these Karnaugh maps will give the required J and K equations for the circuit.

The first state, $Q4 \cdot Q3 \cdot Q2 \cdot Q1$ changes from 0000 to 0001. Applying the rules described in Table 6.3, $Q4$, $Q3$ and $Q2$ need to maintain a 0 and $Q1$ has to toggle from a 0 to a 1. The connections to $J1$ and $K1$ therefore become '0 and x', while those for $J2$ and $K2$, $J3$ and $K3$ and $J4$ and $K4$ become '1 and x'. This is repeated for all the required state transitions and Table 6.6 is obtained. The Karnaugh map for J_1 is then drawn as shown in Figure 6.23, and $J_1 = 1$

Table 6.6

Initial State $Q4 \cdot Q3 \cdot Q2 \cdot Q1$	Next State $Q4 \cdot Q3 \cdot Q2 \cdot Q1$	Inputs			
		J4K4	J3K3	J2K2	J1K1
0000	0001	0x	0x	0x	1x
0001	0010	0x	0x	1x	x1
0010	0011	0x	0x	x0	1x
0011	0100	0x	1x	x1	x1
0100	0101	0x	x0	0x	1x
0101	0110	0x	x0	1x	x1
0110	0111	0x	x0	x0	1x
0111	1000	1x	x1	x1	x1
1000	1001	x0	0x	0x	1x
1001	0000	x1	0x	0x	x1

Repeating for the other columns, the Karnaugh maps for K_1, J_2, K_2, J_3, K_3, J_4 and K_4 are then drawn as shown in Figure 6.24, and the following equations are obtained:

$$J_1 = K_1 = 1$$

$$J_2 = Q_1 \cdot NQ_4$$

$$K_2 = Q_1$$

J_1	$\bar{Q_4}\cdot\bar{Q_3}$ 00	$\bar{Q_4}\cdot Q_3$ 01	$Q_4\cdot Q_3$ 11	$Q_4\cdot\bar{Q_3}$ 10
$\bar{Q_2}\cdot\bar{Q_1}$ 00	1	1	X	1
$\bar{Q_2}\cdot Q_1$ 01	X	X	X	X
$Q_2\cdot Q_1$ 11	X	X	X	X
$Q_2\cdot\bar{Q_1}$ 10	1	1	X	X

Figure 6.23

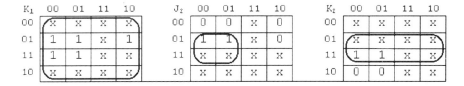

K_1	00	01	11	10
00	X	X	X	X
01	1	1	X	1
11	1	1	X	X
10	X	X	X	X

J_2	00	01	11	10
00	0	0	X	0
01	1	1	X	0
11	X	X	X	X
10	X	X	X	X

K_2	00	01	11	10
00	X	X	X	X
01	X	X	X	X
11	1	1	X	X
10	0	0	X	X

J_3	00	01	11	10
00	0	X	X	0
01	0	X	X	0
11	1	X	X	X
10	0	X	X	X

K_3	00	01	11	10
00	X	0	X	X
01	X	0	X	X
11	X	1	X	X
10	X	0	X	X

J_4	00	01	11	10
00	0	0	X	X
01	0	0	X	X
11	0	1	X	X
10	0	0	X	X

K_4	00	01	11	10
00	X	X	X	0
01	X	X	X	1
11	X	X	X	X
10	X	X	X	X

Figure 6.24

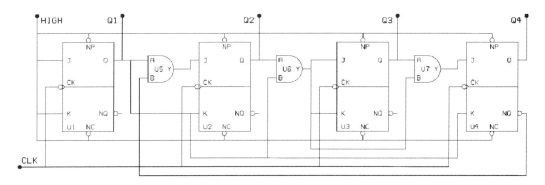

Figure 6.25

$$J_3 = Q_1 \cdot Q_2$$

$$K_3 = Q_1 \cdot Q_2 = J_3$$

$$J_4 = Q_1 \cdot Q_2 \cdot Q_3$$

$$K_4 = Q_1$$

Implementing these equations we get the circuit shown in Figure 6.25. Save it as S_DECADE.SCH and simulate in PULSAR. Using a 10 kHz clock input, you will observe that the glitches have disappeared. It is now possible to achieve very high

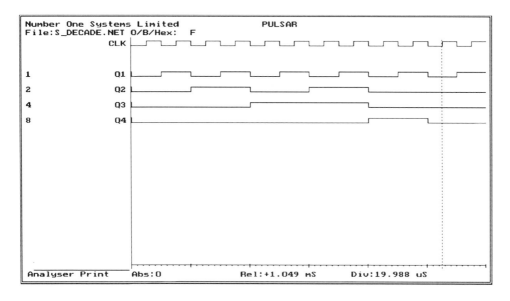

Figure 6.26

speeds without too much trouble (Figure 6.26). Set the Abs reference to 0 and zoom in until you reach a scale of 312 ns/div. Apply a 1 MHz clock. You should just about observe the effects of the JK propagation delays. Account for the 20 nS lag on the output of Q4.

It is interesting, but not very helpful, to examine the effects of an even higher speed clock on this circuit. Take the PULSAR scale down to 39 ns and apply a 10 MHz clock. The limitations of your JK primitives are now quite obvious.

Synchronous 8421-BCD (down) counter with feedback

Using the same approach as used above, we obtain Table 6.7. The Karnaugh maps shown in Figure 6.27 are then obtained. You should obtain the following equations

Table 6.7

Initial State $Q4 \cdot Q3 \cdot Q2 \cdot Q1$	Next State $Q4 \cdot Q3 \cdot Q2 \cdot Q1$	Inputs			
		J4K4	J3K3	J2K2	J1K1
0000	1001	1x	0x	0x	1x
0001	0000	0x	0x	0x	x1
0010	0001	0x	0x	x1	1x
0011	0010	0x	0x	x0	x1
0100	0011	0x	x1	1x	1x
0101	0100	0x	x0	0x	x1
0110	0101	0x	x0	x1	1x
0111	0110	0x	x0	x0	x1
1000	0111	x1	1x	1x	1x
1001	1000	x0	0x	0x	x1

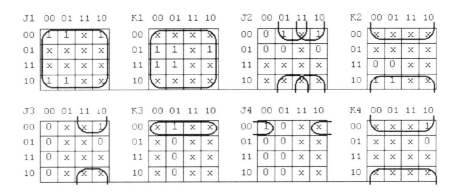

Figure 6.27

from the Karnaugh maps:

$$J_1 = K_1 = 1$$

$$J_2 = NQ_1 \cdot Q_4 + NQ_1 \cdot Q_3$$

$$K_2 = NQ_1 = K_4$$

$$J_3 = NQ_1 \cdot Q_4$$

$$K_3 = NQ_1 \cdot NQ_2$$

$$J_4 = NQ_1 \cdot NQ_2 \cdot NQ_3$$

$$K_4 = NQ_1$$

All that remains is for you to draw the circuit and simulate to verify the equations.

Practical exercise: printing out schematic and timing diagrams

Having got this far with your studies you are probably already contemplating 'hard copy' records of some of the circuits and other diagrams that you have developed. In this exercise we will examine how to print out EASY-PC Pro and PULSAR diagrams.

A variety of output devices are supported by both packages, and it is assumed that the printer of your choice, be it parallel, serial or even a plotter, has already been installed and works properly. If you have any difficulties, consult your device guide, DOS manual or, if you have the full version, Part 5 of the EASY-PC Pro manual.

EASY-PC Professional XM

We will take the first counter circuit, the asynchronous 4-bit natural binary counter, as our working example and print it out as shown in Figure 6.1. You will recall that the drawing area available in EASY-PC Pro is rather large. Because you are probably going to want to print to a standard printer using a standard paper size (typically A4), it is important to select the area of your diagram first. The following steps describe the procedure:

Step 1: Draw a block around the circuit: SHIFT C increases the size of the cursor and thus makes accurate framing easier (Figure 6.28).

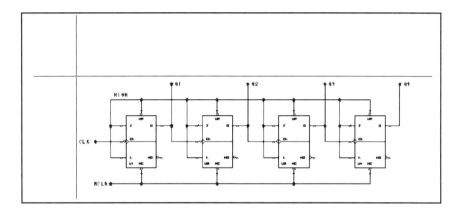

Figure 6.28

Step 2: Start from the top left-hand corner and press F10 to generate a block.
Step 3: Move the cursor diagonally across the diagram area to the bottom right, making sure that the part that you want is contained within the block (Figure 6.29).

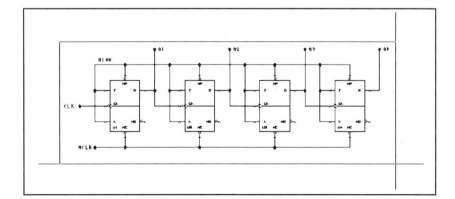

Figure 6.29

Continued on p. 144

Practical exercise: printing out schematic and timing diagrams *(Continued)*

Step 4: Click the left-hand mouse button, or press ENTER and the selected area will be highlighted with the cursor ending up at the bottom left corner of the block (Figure 6.30).

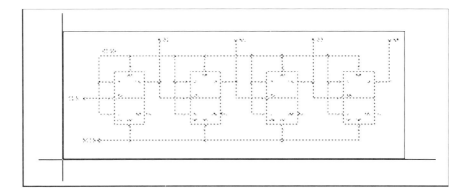

Figure 6.30

Step 5: Click on the Output menu at the top of the screen to reveal output options (Figure 6.31).

```
Output

    Output to
Pen-Plot
Dot-Matrix
Laser Jet II
Parts List
```

Figure 6.31

You will see that EASY-PC Pro generates output in a whole variety of formats:

- Pen-Plot is for Hewlett Packard plotters that require HPGL-A and HPGL-B formats.
- Dot Matrix covers all Epson and IBM graphics compatible printers with either 9- or 24-pin heads.
- Laser Jet II is the option for all Hewlett Packard compatible laser and ink jet printers.

The remaining option on this menu is a feature that will be covered elsewhere. The similar menu which appears in the PCB mode also contains options to output to Gerber and NC drill files (Excellon format), and DXF on EASY-PC Pro XM only.

Step 6:   Select the best option to suit your installation. EASY-PC Pro generates a temporary file 'TEMPFILE.SCH' and switches to a new screen (Figure 6.32).

```
           EASY-PC Professional, Laser printer output
   Layer
 0 Silk       On                     I Input From      : TMPFILE.SCH
 1 Copper     On   Resist (.LUx)     O Output To       : LPT1
 2 Copper     Off                    L Layers output   : Together
 3 Copper     Off                   ▲O Solder resist   : No
 4 Copper     Off                    H Pad Holes       : Avoid
 5 Copper     Off                    M Resolution      : 300 dpi
 6 Copper     Off                    P Paper           : A4
 7 Copper     Off                    N Copies          : 1
 8 Copper     Off                   ▲E Scale           : 1.000
 9 Copper     Off                   ▲F Print from      : 10.600,19.550  in
 A Copper     Off                    T Print to        : 17.300,21.950  in
 B Copper     Off                   ▲P Print offset    : 0.547,4.447    in
 C Copper     Off                    G Pin names       : On
 D Copper     Off                   ▲G Pin numbers     : On
 E Copper     On                    ▲A Compensation ▲I Summary
 F Silk       Off                   ▲C Centre Print   S Start Print
                                    ▲S Save Setup     R Restore Setup
   Print will FIT                    U Units          Q Quit
```

Figure 6.32

There is quite a lot of information on this screen, but fortunately EASY-PC Pro has made some decisions first and all you have to do is check a few details.

Step 7:   First check that your diagram *fits* into the print area. It should do if you stuck to the dimensions suggested when designing the primitive component and selected a reasonably tight outline or area to be printed.

Step 8:   If the print does not fit, try reducing the scale by 0.05 to 0.95 by clicking on the entry, or typing [SHIFT] [E].

Step 9:   Centre the print when you have done this: [SHIFT] [C] for the laser option.

Step 10:   Repeat step 8 until Print will FIT.

Continued on p. 146

Practical exercise: printing out schematic and timing diagrams (*Continued*)

Step 11: Schematic drawings are usually done on layers 0, 1 and sometimes E (together).Make sure that these layers are 'on'. (Lines use layer 0, components layer 1, and flipped components layer E.)

Step 12: Option L is forced to the value together for schematics. Type `L` to confirm this. You will need to set it only if you are printing a layout.

Step 13: Decide whether you want to display either pin numbers, pin names, or both by clicking on or typing `G` or `SHIFT` `G`. This is just down to personal preference.

Step 14: Make sure that your output is directed to the correct output port: LPT1, COM1, File, etc. You set this with option `O`. If you choose File the file extension is automatically set.

Step 15: Check that the print resolution is compatible with your printer: e.g. 300 dpi for an ink-jet printer. This is option `M`.

Step 16: To avoid having to go through all this process again, save this current configuration by clicking on or typing `SHIFT` `S`.

Step 17: Now you are ready to print. Centre the print, if you have not done so already. (The hot key depends on the output type selected.)

Step 18: Start the print by clicking on or typing `S`.

Step 19: When printing has finished you can either quit back to EASY-PC Pro schematic by typing `Q`, `Y` or you can repeat the print with different settings if the current output is unsatisfactory.

Back in EASY-PC Pro you are presented with your original schematic diagram still in a block highlight. This can be removed by selecting Group Ops from the EDIT menu, starting a group, and then pressing `ESC`; or by pressing `F10` `ESC` (Figure 6.33). You can now make any alterations (such as changing the position or size of text items).

File	Edit	View
Edit Line	F1	
New Line	F2	
Edit Text	F5	
New Text	F6	
Edit Component	F7	
New Component	F8	
Group Ops	F10	

Figure 6.33

You may wish to print out a larger image. This can be done by rotating the selected block through 90°: `A` `1`. Rotation will be about the bottom left-hand corner, so make sure that the image does not go outside the limits of the drawing area. If it does stray outside, either shift the block to the right first until rotation succeeds, or define a new relative origin in the middle with �. The group will always rotate about the relative origin. Back in the printer output menu, try increasing the scale to 1:1.35, and then repeat steps 9 and 10 before continuing.

PULSAR

Printing out your PULSAR screen is a much easier process. But before you start it is important that the package is set up to recognize your particular printer and the port to which it is attached.

Step 20: Click on Configuration in the top menu bar to reveal these options (Figure 6.34).

```
Configuration
General Settings
Path Settings
File Extensions
Analyser Settings
Equipment
Copyright Message
Revision
Leave the Menu
```

Figure 6.34

Step 21: Choose Equipment to find the Printer Type and Printer Destination settings (Figure 6.35).

```
Equipment
Printer Type
Printer Destination
Set Up Serial Ports
Mouse Gearing
Reverse Mouse Buttons
Keyboard Gearing
Leave the Menu
```

Figure 6.35

Step 22: Click on Printer Type and you now have a choice of four options (Figure 6.36).

Continued on p. 148

Practical exercise: printing out schematic and timing diagrams (*Continued*)

```
Printer Type
9/24 pin Dot Matrix
Alt 9 pin Dot Matrix
Laserjet II
GEM IMG File Format
Leave the Menu
```

Figure 6.36

Step 23: Now it is just a matter of selecting the most appropriate version – the comments made in step 5 apply. A useful option here is to generate an output file that can be imported into a report or other document, a facility which has been used extensively throughout this book! The file will be in the GEM IMG format, but this can be converted into a variety of other types by using any of a host of graphics utilities.

Step 24: To complete the set-up you will need to specify the port to which your printer is connected. Leave the Printer Type menu and choose Printer Destination (Figure 6.37).

```
Printer Destination
LPT1:
LPT2:
LPT3:
COM1:
COM2:
File
Leave the Menu
```

Figure 6.37

Step 25: Select the appropriate destination and leave this menu. As you leave the configuration menu be sure to Save the Settings before returning to the main PULSAR screen.

Step 26: Printing is now a matter of clicking on File for Analyser Operations and choosing Print the Screen (Figure 6.38).

```
Analyser Operations
Load a Circuit
Create/Modify a Circuit
Delay Modifiers
Print the Screen
Re-Build the Circuit
Re-Simulate the Circuit
Help
Leave the Menu
```

Figure 6.38

Exercises

1. Investigate the operation of the following counter circuit. Is it asynchronous or synchronous? What is its modulus?

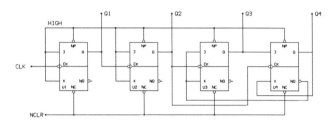

Figure Ex6.1

2. Devise a 4-bit synchronous variable modulus counter that stops on completion of its count. Could the circuit be modified so that the count is continuous?

3. For nonbinary counters, other count sequences are easily obtained by using the 'feedback' technique. Devise a 4-bit synchronous Gray code counter.

4. *Integrated assignment idea (for BTEC)*: With the 'feedback' technique in mind, use a spreadsheet to create a model that can generate Karnaugh maps for any:

   (i.) 4-bit count sequence and

   (ii.) 5-bit count sequence.

   Use this model to generate a synchronous 4-bit modulo-9 down counter. Check the Gray code counter you devised in Exercise 3. Could this technique be expanded to cover the design of a seven-segment display code counter?

7 Registers

Introduction

In the last chapter we saw how JK flip-flop elements connected in cascade could be used to create a whole variety of counter circuits. Yet another application for flip-flops is for storing bits of information.

We saw earlier that the D-type arrangement was able to store a single bit of information. Several such flip-flops connected in cascade can therefore be configured to store multi-bit information. Such circuits are referred to as *registers*. Whereas counters are used to keep track of a sequence of events, registers are used to store and manipulate data that contribute to all or many of these events.

In computer control systems it is necessary to hold data words for short periods of time. The number of bits required to represent a machine instruction is termed a *machine word* (or *word* for short). Each word can represent not just a machine instruction, but a numerical value or a character code as well. A word can be 4, 8, 16 or 32 bits wide depending upon the type of controller being used. A register then is a group of bistable elements grouped together in 4s, 8s, 16s, 32s etc., but which act as a single unit to hold a machine word or data, or to perform certain functions on it. For example, in an 8-bit system an 8-bit register would not only hold a byte of information but can also be made to decrement (subtract 1), increment (add 1), shift left (multiply by 2^n) or shift right (divide by 2^n).

Registers are classified according to the way information bits are stored and retrieved. If data are stored and removed from either end of a multi-bit register, one bit at a time, the register is referred to as a *serial or shift register*. However, if all bits of the word are stored or retrieved simultaneously, the register is referred to as a *parallel or buffer register*. Buffer registers principally hold information, whereas shift registers generate predictable sequences, shift signals in time and convert a serial bit stream into parallel form (or vice versa).

In this chapter we will develop each of the principal register types and consider some applications.

The 4-bit serial (shift) register

As its name implies, the shift register is used primarily to move data about in a digital system. A shift register collects each bit as it is entered at one end of the register, then shifts it along into more significant places as new digits are entered. In Figure 7.1 there are four D-type bistables; each output is connected to the next stages D input, with each stage having a common clock. The clock in this case is merely a counter which counts in all the relevant bits, and then stops.

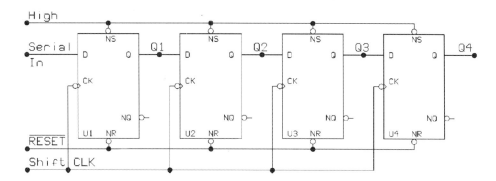

Figure 7.1

All bistables must be re-set initially so that all outputs start at logic 0, and then a 4-bit input is applied to D1. A clock pulse causes the first bit to appear at the output of the first stage, Q1. This then becomes the input to the second stage and at the next clock pulse it moves on to Q2 whilst a new bit appears on Q1. The sequence continues until each of the four data bits are latched into each of the four bistables. This gives us serial input; the output can be shifted out as serial data at Q4 which gives us a 'serial-in serial-out', or SISO, register. Alternatively, the output can be taken as parallel data from Q1 to Q4.

One application of a SISO register is to delay digital signals. The length of delay depends upon the clock frequency and the number of bistables in the circuit. Certain integrated circuits contain shift registers comprising 4096 flip-flops. Such a device may be used to provide a 'reverb' system where an analogue audio signal is converted into digital form, delayed by the shift register and converted back into its analogue form again.

Exercise

Construct the circuit shown in Figure 7.1 using the DT7474 primitive flip-flop element you created in Chapter 5, and simulate it in PULSAR.

Use Clear on the NR input to re-set the bistables, then apply a suitable data stream as input after the Clear has been negated. Any 4-bit data value will do; this example uses 1001.GEN which should be provided on your disk. The clock should be faster than the applied signal; in this case 40 kHz. Note the 87.5 µs delay between the leading edge of the applied data and its appearance on the output of Q4.

Continued on p. 152

Exercise *(Continued)*

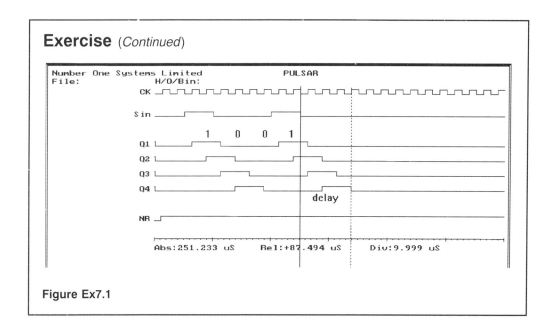

Figure Ex7.1

The 4-bit parallel (buffer) register

This buffer register receives data on each of its four bistables at the same time (known as *parallel loading*) and on the application of a trigger pulse transfers the data to the four outputs. We have a 'parallel-in, parallel-out', or PIPO, register.

As shown in Figure 7.2, again we have four D-type bistables. This time they are not connected in cascade but each input is applied to its corresponding D input, with each output taken from Q. There is a common clock which just acts as a simple trigger or enable pulse. All bistables must be re-set initially so that all outputs start at logic 0, then a 4-bit input is applied. A trigger pulse latches each of the four data bits into each of the four bistables, (Figure 7.3). These circuits are used as temporary stores and for data manipulation.

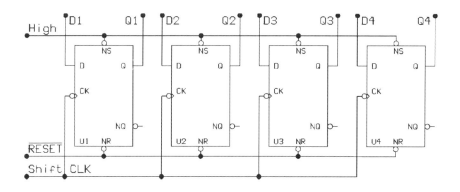

Figure 7.2

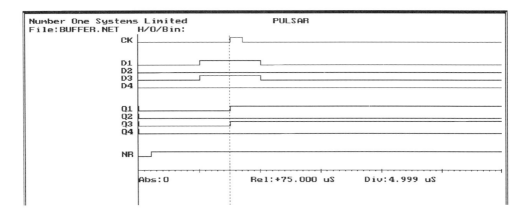

Figure 7.3

Exercise

Use Clear on NR to re-set the bistables. Create generators for the data and clock so that, after 50 μs, high pulses of 50 μs duration are applied to D1 and D3 to represent 1010_2. CK has to arrive after data have been applied, so that on the positive edge of a 10 μs clock trigger pulse data appear on the four output pins.

The 4-bit universal shift register

With a little refinement we can combine the parallel and serial operations to give us a 'parallel-in, serial-out', or PISO, circuit, which is very useful for data conversion (Figure 7.4). The data are again applied to the PRESET inputs via some gating circuits. Upon receiving an Enable signal, normally the first shift pulse, the data are gated through to PRESET. The next shift pulse transfers PRESET input to the Q outputs, where they can be read off in parallel again or be shifted on through to Q4 thus providing a serial output. We can apply the same gating techniques to the parallel outputs (Figure 7.5).

In both cases parallel data can be presented to the input or output when required, without loading the serial channel. Combining the two circuits will give us the far more practical universal shift register which provides all four modes of operation (Figure 7.6):

- serial-in, serial-out (SISO),
- parallel-in, parallel-out (PIPO),
- serial-in, parallel-out (SIPO) and
- parallel-in, serial-out (PISO).

Parallel operation (PIPO and PISO)

When a high or logic-1 signal is applied to Parallel IN Enable, any data present on the four parallel input lines will be presented to the four bistable PRESET inputs via

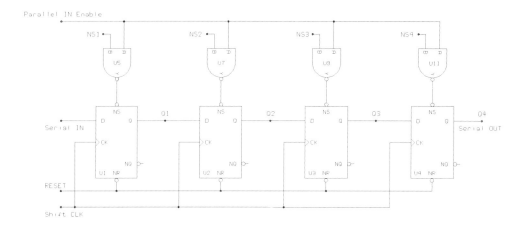

Figure 7.4

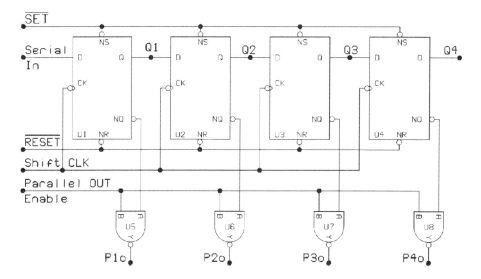

Figure 7.5

NAND gates as complemented data. Upon the first shift clock, data will appear latched in its original form at each bistable output, by which time the enable signal should have been removed. The data are now held in store.

One of two processes can now happen:

- The data are presented as parallel output upon receiving a logic-1 Parallel OUT Enable pulse.
- The data are shifted out at Q4 (or Serial-out) upon the application of three shift pulses. Note that Serial-in should be tied low if the bistables are Cleared at the outset. A fourth clock pulse would clear the output to low.

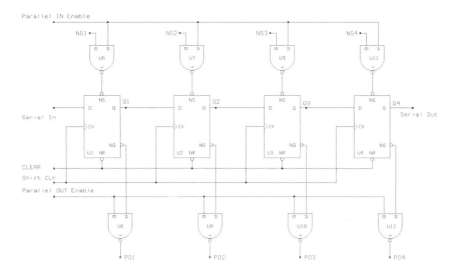

Figure 7.6

Serial operation (SISO and SIPO)

Now the Parallel IN Enable signal should be disabled and the bistables Cleared. Serial data are applied to the Serial-in, LSB first, and four shift clocks will load the word into the register where it will be held. Then, either:

- three more shift clocks will shift the data out at Serial-out or
- data will be presented to the parallel output upon the assertion of a Parallel OUT Enable pulse.

Typical applications are as follows:

- SISO – for delay circuits; filtering; sequence generation (counting).
- PIPO – as a buffer or temporary store; for fast movement of data within a micro-computer system.
- PISO – for converting parallel data to serial data in order to allow transmission along a single signal channel, e.g. in modems that drive telephone lines or radio links.
- SIPO – for converting serial data back to parallel form.

Each of the four modes is examined in the Practical Exercise at the end of this chapter.

The 4-bit shift left register

Apart from these hardware applications, the shift register can also be used to perform arithmetic functions. As stated earlier in this chapter, shifting data one place to the right effectively divides it by 2, two places by 4, and so on. So our shift register can be used as a divider. A shift left register will multiply the data by 2 for each shift. We can modify the circuit shown in Figure 7.1 to achieve this (Figure 7.7).

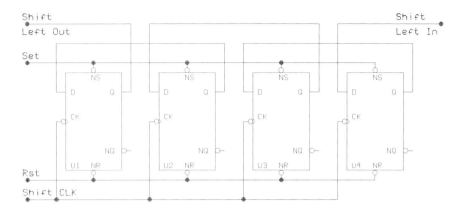

Figure 7.7

The 4-bit bi-directional shift register

This is a much more practical register, which combines both the shift left and right operations (Figure 7.8). The serial signal channels are the same as those shown in Figures 7.1 and 7.7, and direction is selected by additional logic similar to that used in Figure 6.18.

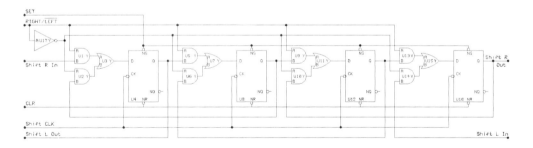

Figure 7.8

To shift right, a logic-1 signal is applied to the direction control line. This enables the top AND gate in each group of selection logic and data and taken from the preceding bistable's output. Serial data and shifted out at the right-hand side.

To shift left, a logic 0 is applied to the direction control line, which enables the bottom row of AND gates. This time data is taken from the output of the next bistable down the chain. In this way data applied at the right-hand side of the circuit are shifted out at the left.

By adding parallel input and output circuitry to the circuit shown in Figure 7.8 we end up with a multi-mode universal shift register. Try it if you wish! It would make an ideal addition to your primitive circuit collection. If you are using the supplied software, the circuit in Figure 7.8 will exceed the limits. Use 2X2ANDOR to replace the grouped gates, as shown in Figure 7.9.

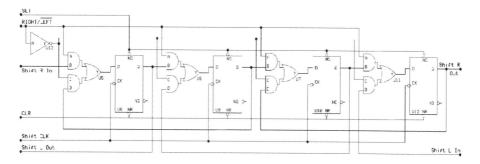

Figure 7.9

Like the programmable counters and the universal shift register, such circuits are ideal candidates for integrated circuit design, and we will tackle some of these in Chapter 8.

Shift registers as counters

Another application of the shift register is to produce counts or controlled sequences by feeding back some of the output to the input. Shift register integrated circuits are used extensively in random bit generators, multiple address coding circuits and parity bit generators.

The ring counter

This is the simplest of the feedback shift registers, where the output from the final stage is fed back directly to the serial input forming a data ring (Figure 7.10). Suppose the register had been PRESET to hold 10000_2. On the first clock pulse a logic 1 would appear at the Q1 output, and the logic 0 on Q5 now provides the input to D1. The next clock pulse shifts all the bits to the right. With this particular data input we get a count table like this:

Count	Q_1	Q_2	Q_3	Q_4	Q_5
1	1	0	0	0	0
2	0	1	0	0	0
3	0	0	1	0	0
4	0	0	0	1	0
5	0	0	0	0	1
6	1	0	0	0	0

A pattern of five unique states is generated using a very simple code – hence a modulo-5 counter. No count decoding is necessary because the code is so simple, the state being indicated solely by the position of the 1 in the table.

Ring counters are synchronous devices and thus have the advantage of high speed operation. They are, of course, much simpler (5 bistables produce a modulus of 5; *n*

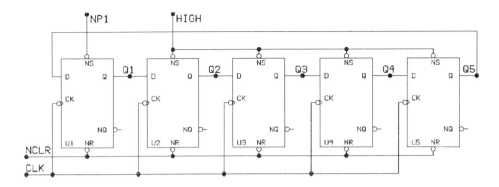

Figure 7.10

bistables produce a modulus of n), but are less efficient than, ripple counters, which have count lengths of 2^n.

A typical application of ring counters is in stepper motor driver circuitry, where pairs of 1s provide the sources for energizing the motor's windings, causing the machine to rotate.

Twisted ring counter

Here the count is doubled. The output from Q_5 is fed back to D_1 through an inverter; alternatively, the output could be taken from $NQ5$ instead (Figure 7.11). This counter has a modulus of 10. In general, n bistables connected in twisted ring will have a modulus of $2n$.

With bistables cleared initially, the count would proceed as follows:

Count	Q_1	Q_2	Q_3	Q_4	Q_5	Code
1	0	0	0	0	0	
2	1	0	0	0	0	$A \cdot NB$
3	1	1	0	0	0	$B \cdot ND$
4	1	1	1	0	0	$C \cdot ND$
5	1	1	1	1	0	$D \cdot NE$
6	1	1	1	1	1	$E \cdot A$
7	0	1	1	1	1	$NA \cdot B$
8	0	0	1	1	1	$NB \cdot C$
9	0	0	0	1	1	$NC \cdot D$
10	0	0	0	0	1	$ND \cdot E$
11	0	0	0	0	0	$NE \cdot NA$

We now have ten different states producing a count known as the *Johnson code* – another name for the twisted ring counter is the *Johnson counter*.

To appreciate the code column in the above table, merely change Q_1 to A, Q_2 to B and so on to Q_5 to E. Start from state 2 and take the logic 1 and its adjacent 0 to derive $A \cdot NB$. We choose this because it is the only $1 > 0$ transition for that state. Code each

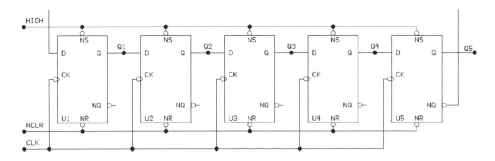

Figure 7.11

state from its leading logic 1 until you reach state 7, then you have to use the 0 > 1 transition. If you follow the sequence through to state 11, where there is no transition, the only code not used is NE · NA; this becomes the code for state 1.

This circuit has all the advantages of a synchronous circuit It has fewer connections, and there is no PRESET logic because all the bistables are cleared initially. It is also more economical than the ring counter, having twice the modulus of a ring counter of similar size. The disadvantages are that it is necessary to decode the output logic, and that only even moduli are possible.

The chain code generator

These counters produce a larger sequence of possible output combinations than other ring counters. Here an Exclusive-OR function provides feedback paths to the serial input, and for n bistables a code cycle length of $(2^n) - 1$ is possible. For the four-stage ring counter shown in Figure 7.12, the outputs from Q4 and Q3 are Exclusive-ORed together and fed back to the serial input.

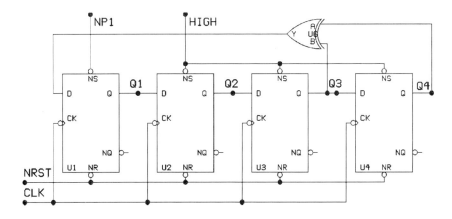

Figure 7.12

With the first bistable pre-set to 1 before the first shift pulse, we get the following count sequence:

Count	Q_1	Q_2	Q_3	Q_4
1	1	0	0	0
2	0	1	0	0
3	0	0	1	0
4	1	0	0	1
5	1	1	0	0
6	0	1	1	0
7	1	0	1	1
8	0	1	0	1
9	1	0	1	0
10	1	1	0	1
11	1	1	1	0
12	1	1	1	1
13	0	1	1	1
14	0	0	1	1
15	0	0	0	1

This circuit produces a repetitive count of 15, as $(2^4) - 1 = 15$.

The count is a pseudo-random one and the 0000 state does not occur, as this would terminate the cycle anyway.

In general, for n bistables the maximum sequence length is $(2^n) - 1$ and the number of states depends upon the points from which the feedback is derived. A maximum count of 31 is possible for a five-stage ring counter. You get this when feedback is taken from either Q3 and Q5 or Q2 and Q5. If the feedback were to be taken from Q4 and Q5, then you only get a count of 21. Minor cycle lengths of 7, 3 and 1 are obtained as the feedback points get closer to Q1.

This circuit provides a pseudo-random binary sequence and can be found in test instruments such as signature analysers, where 16 such stages are possible.

Practical exercise: investigating the universal shift register

The objective here is to draw and examine two universal shift registers. To help us with the simulation, this time we will also investigate the generator utility that comes with PULSAR.

The first task is to draw the schematic of the universal shift register shown in Figure 7.6, if you haven't done so already. Save the circuit as U_SHIFT.SCH, making sure that the node names are the same as those given to you in the supplementary diagram in Figure 7.13. If you are using the disk supplied with this book, there are too many NAND gates in this circuit. Instead you will need to use the component 2X2NAND for each pair of gates. The circuit will look like the one shown in Figure 7.14.

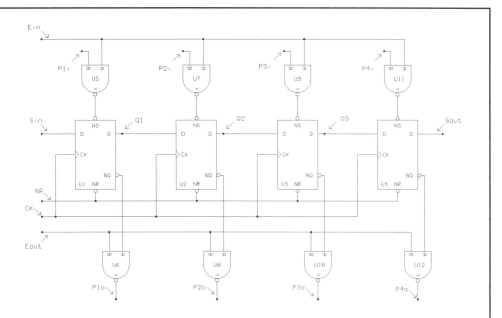

Figure 7.13

Now enter PULSAR by selecting Logic – PULSAR from the Tools menu.
We want to examine each of the four modes of operation:

- serial-in, serial-out (SISO),
- parallel-in, parallel-out (PIPO),
- serial-in, parallel-out (SIPO) and,
- parallel-in, serial-out (PISO).

To do this we must decide upon what signal shapes and control signals we need
to use in order to exercise each of these four functions.

To test the serial-in, serial-out (SISO) mode we will need to create three special
signals: one for the serial data itself, another for the 'shift clock' and a 're-set'
signal. The serial data will be some 4-bit word: this example uses 1001_2.

- The 're-set' signal will need to be a single negative pulse, roughly half the period
  of the shift clock, which itself will need to be at least 9 pulses long.

- One CK for 're-set', four more to shift the 4-bit number into the four bistables
  and another four pulses to shift all the data out at S_{out}, making 9 pulses in all.

- The frequency of the applied clock will be 25 kHz, for no reason other than it will
  give us convenient 20 µs pulse widths to use when we create our generators.

Continued on p. 162

Practical exercise: investigating the universal shift register
(*Continued*)

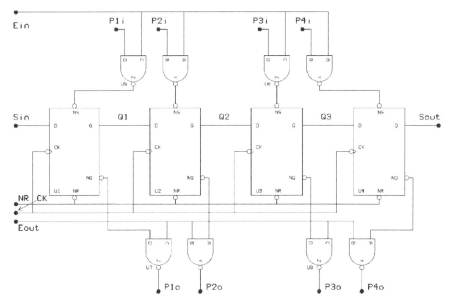

Figure 7.14

The 'shift clock' signal

Step 1:   Click on Generator in the top menu, to give you the Generator Operations menu (Figure 7.15).

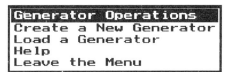

Figure 7.15

Step 2:   Choose Create a New Generator.

This brings up the 'generator edit' screen (Figure 7.16). You will be confronted with:

● a thin horizontal red line, which is the signal path or frame to be edited; and
● a thin vertical blue line, which is the cursor.

Note the bottom menu and the editing keys, as well as the scale. Note also that the absolute cursor is at position zero on this scale.

Figure 7.16

Step 3: Use ⬜ or ⬜ to give your scale a calibration of 10 µs/division; or use ⬜ and type in the figure.

The generator

Step 4: Either click on Ins on the bottom menu, or just key ⬜ to insert a section.

This brings up the state selection menu (Figure 7.17).

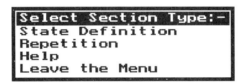

Figure 7.17

Step 5: Choose State Definition (Figure 7.17).
Step 6: Select STRONG LOW (Figure 7.18).

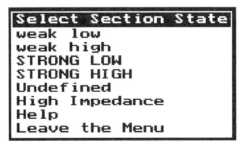

Figure 7.18

Step 7: Give this low section a time duration of 20 µs.

Continued on p. 164

Practical exercise: investigating the universal shift register
(*Continued*)

Notice that the cursor has moved to the right by 20 μs (check the scale reading at the bottom of the screen). The trace has changed colour from red to white and has, in fact, dropped down the screen a fraction to indicate a low level.

Step 8:   Repeat steps 4 to 7 and INS a second section. Make this second section a 'strong high', of 20 μs duration.

Step 9:   Press B (the cursor moves back one section) and then press M. Note that Mark now shows 1. Press E to move the cursor to the end.

Step 10:   Repeat steps 4 to 7 and INS a third section. Make this new section a 'strong low', of 20 μs duration.

The first pulse is now completed and the cursor should be at the 60 μs point on the scale.

Step 11:   Press INS again, but this time select Repetition. From mark 1 enter four repetitions, and four more pulses will appear. The cursor should be at 220 μs (Figure 7.19).

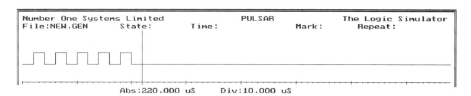

Figure 7.19

Step 12:   Save this generator by clicking on File to bring up the Generator Operations menu and choose Save the Generator (Figure 7.20).

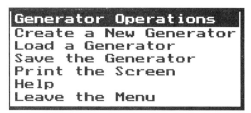

Figure 7.20

Step 13:   Save the generator as 5PULSES.GEN. Note that the filename given in the top PULSAR menu has changed from NEW.GEN.

Step 14: Add a sixth pulse. Press ⬚B and click on Repeat. Change to five repeats, and press ⬚E. The cursor should be at 260 μs. Save this as 6PULSES.GEN.

Step 15: Repeat step 14 with eight repeats to complete a ninth pulse − the cursor should have reached 380 μs (Figure 7.21). Save this as 9PULSES.GEN.

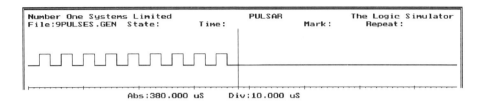

Figure 7.21

Use ⬚B and ⬚N to move back and forth along the pulse trace and check each section. Note the State and Time display for each section. These can be altered for any section at any time by keying either ⬚S or ⬚T.

Move back to the first pulse (abs @ 20 μs) by using ⬚HOME or ⬚H then ⬚N ext and key ⬚S (for state). Change the state to High Impedance and observe the result (Figure 7.22).

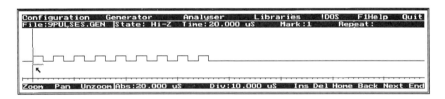

Figure 7.22

Change the state back to Strong High, and then alter the time for that section to 40 μs (Figure 7.23): ⬚T ⬚HOME ⬚4 ⬚DEL ⬚ENTER . Restore the period for that pulse to 20 μs with ⬚T ⬚HOME ⬚2 ⬚DEL ⬚ENTER .

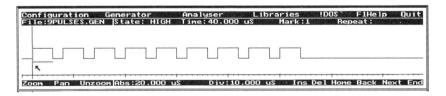

Figure 7.23

Continued on p. 166

Practical exercise: investigating the universal shift register
(*Continued*)

You should now be ready to tackle the other generators which are needed to test your universal shift registers.

The serial input data

To represent 1001_2, make the first section a Strong Low of duration $45\,\mu s$; follow this with a STRONG HIGH section of duration $40\,\mu s$; a Strong Low section of $80\,\mu s$; another Strong High section of $40\,\mu s$, and ending with a Strong Low section. Note that the timing of the final state is not vital, because all you need to do is return a permanent Low, although the value chosen ought to be of the same order of magnitude as those already used. A more convenient short-hand for this sequence would be:

Low, $45\,\mu s$, High, $40\,\mu s$, Low, $80\,\mu s$, High, $40\,\mu s$, Low, $40\,\mu s$.

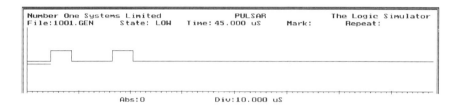

Figure 7.24

Save this generator as 1001.GEN.

A $10\,\mu s$ clear signal

Create this signal (Figure 7.25) using the sequence:

Low, $10\,\mu s$, High, $100\,\mu s$,

and save as CLR_10US.GEN.

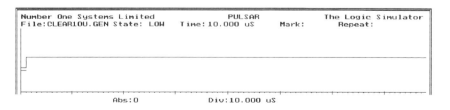

Figure 7.25

A parallel input enable pulse

Create this pulse, (Figure 7.26) using the sequence:

Low, 60 μs, High, 10 μs, Low, 250 μS

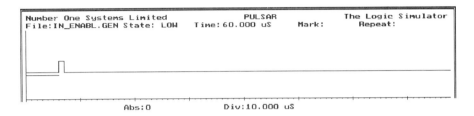

Figure 7.26

and save as IN_ENABL.GEN.

A parallel output enable pulse

Create this pulse (Figure 7.27) with the sequence:

Low, 200 μs, High, 10 μs, Low, 50 μs

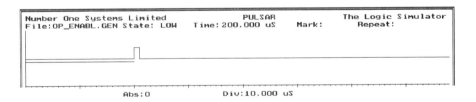

Figure 7.27

and save as OP_ENABL.GEN.

A single 50 μs pulse

This serves as a logic 1 for the parallel input. Create the pulse (Figure 7.28) using the sequence:

Low, 50 μs, High, 50 μs, Low, 100 μs

Continued on p. 168

Practical exercise: investigating the universal shift register
(*Continued*)

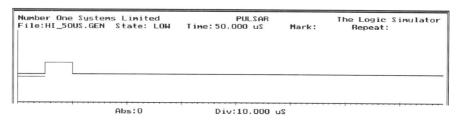

Figure 7.28

and save as HI_50US.GEN.

Step 16:   Leave Generator by clicking on Analyser in the top menu. This should return you to the simulator screen, which shows your unformed signal traces.

Step 17:   The next step is to arrange the traces into a logical order before we apply the six generators just created.

Simulation

We want to test each mode of our register in turn.

Starting with the SISO, arrange the traces in the order shown in the timing diagram in Figure 7.29. Push the others to the bottom of the screen, as they are not needed for this first test, making sure that they are disabled before you do so. Apply 9PULSES.GEN to the CK input; 1001.GEN to S_{in}, CLR_10US.GEN to NR, and LOW.GEN to E_{in}. Observe that the bistables are cleared during T1 (the first CK pulse) and that it takes eight more T states or clocks to shift the 1001 signal out at S_{out}.

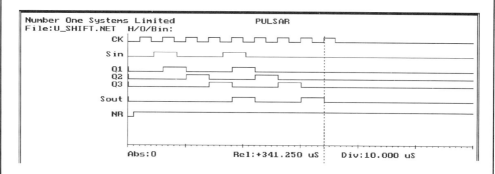

Figure 7.29

Moving on to SIPO, we need to bring E_{out} and the four parallel output lines into play (Figure 7.30). This time it only takes a clear and four pulses to present the data to the parallel outputs; so change the CK input to 5PULSES.GEN. Apply OP_ENABL.GEN to E_{out}; this should coincide with the fifth CK pulse. Line up the relative cursor with the trailing edge of the applied signal and observe the binary display. Remember to turn off the unwanted signals in the Weightings column first.

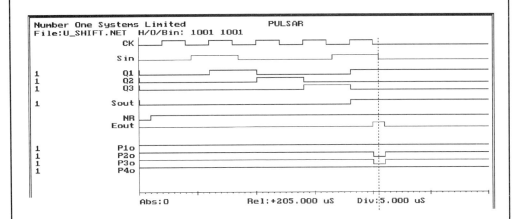

Figure 7.30

We will need to reorganize the display a little in order to test the two parallel modes. For PISO, retrieve the parallel input enable, E_{in} and the four parallel input traces (P1i, P2i, etc.) (Figure 7.31). You won't be needing E_{out} and P_{out} for this test. Clear the 1001 signal from S_{in} by replacing it with LOW.GEN, making P2i and

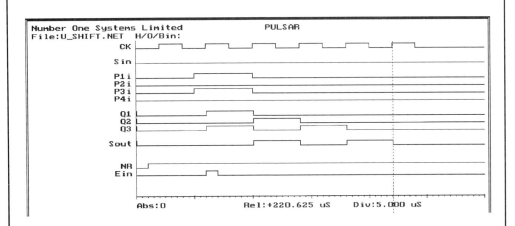

Figure 7.31

Continued on p. 170

Practical exercise: investigating the universal shift register
(*Continued*)

P4i low while you're about it. Then apply your HI_50US.GEN to P1i and P3i and IN_ENABL.GEN to E_{in}. To simulate you will need to use 6PULSES.GEN on CK, T1 for Clear, T2 to latch the input enable, and T3 to T6 to shift out the data on S_{out}.

Finally, we have the PIPO mode. Move S_{in} out of the way, but keep a Low on it. Bring back the output enable and the four parallel output traces (Figure 7.32). All we need is a single clock to coincide with IN_ENABL.GEN; you can use the same signal on CK as E_{in} but keep the same parallel input data. Any time you apply your output enable signal, parallel data will appear on the P_o lines. Check that the value is correct by monitoring it with the corresponding cursor.

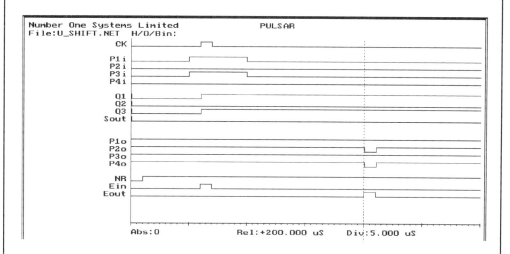

Figure 7.32

Part 4

8 Logic Circuit Families

Introduction

Progress in electronics has resulted in the miniaturization of equipment and components. Besides a significant reduction in size, which has the obvious benefit of space saving, these smaller circuits now make less of a demand on power sources and thus produce more economic and reliable systems. Miniaturized circuit packages may broadly be divided into three groups: film circuits, hybrid circuits and monolithic integrated circuits (the latter being generally known as ICs). Let us briefly consider each of these three.

Film circuits are constructed by depositing films of conductive and nonconductive materials on some insulating base or substrate. All passive component types are possible with high values of component accuracy. Resistors are usually single conductive films. Capacitors of up to 5000 pF are constructed using film layers in 'plate' form, and inductances of a few microhenries are constructed by depositing material in spiral form. MOSFETs may also be constructed, but only in small numbers. Film circuits may be further subdivided into *thin film* or *thick film* types, where the relative thicknesses of the material are $10^{-6}-10^{-4}$ and $10^{-4}-10^{-2}$ inches, respectively.

Hybrid circuits combine the accuracy of the component values of film circuits with the circuit complexities of monolithic integrated circuits.

Monolithic integrated circuits are by far the most popular and familiar of the miniaturized devices. They duplicate discrete logic circuits by constructing them on single silicon slices or chips; hence the term 'monolith', which is derived from the Greek and means 'single stone'. Resistors, capacitors, diodes and transistors may be constructed by layering various semiconductor materials on top of one another. Coils are not included in this list because the value of the inductance down at chip size is so small as to be useless.

Circuit complexity

Initially, simple logic gate circuits were constructed on a single silicon chip, but it soon became clear that combinational logic circuits could be built onto the same chip at practically no extra cost. Larger and more complex electronic circuits could be fabricated on larger areas of silicon, while today, electronic circuits comprising millions of components can be built onto a single silicon chip. This gives rise to a need for classifying the complexity levels of IC types. These have become known as 'scales of integration' and they grade monolithic ICs roughly into groupings according to the number of gates they contain.

- SSI (small scale integration) has up to 10 gates on a single IC. For example, six inverters or four AND gates.
- MSI (medium scale integration) has between 10 and 100 gates on a chip. Typical examples are counters and registers.
- LSI (large scale integration) has between 100 and 10 000 gates on a chip. This large grouping includes some memory devices, programmable input/output and 8-bit controllers.
- VLSI (very-large scale integration) has between 10 000 and 100 000 gates on a chip. We are now in the realm of 16-bit microprocessors and programmable logic arrays (PLAs).
- ULSI or GSI (ultra-large, or grand, scale of integration) covers higher complexities still. A single chip in this group will contain more than 100 000 gates. The commonest devices in this area are leading edge memory chips.

Logic circuit requirements

Before considering some of the current technologies, it is important to identify those properties that a digital logic circuit ought to possess.

High speed. This is an important consideration. Digital circuits have to perform a large number of operations every second. They have to approach a 'real-time' situation where control and instrumentation process signals almost instantaneously. Propagation delays are an important consideration here; the greater the propagation delay the lower the maximum frequency. Devices that have delays of the order of 10 ns will experience problems as 100 MHz clock speeds are approached. Usually the penalty for speed is high power consumption.

Low power dissipation. A large number of devices is usually needed to perform complex functions. Therefore each active device should dissipate minimum amounts of power to keep down total power consumption and consequent rises in temperature. Furthermore, equipment may have to run from battery supplies and thus should be economical to operate. The 'speed–power' product gives a basis for comparing logic circuits where propagation delay and power dissipation are important considerations. The lower the product, the better.

High packing density. It would be distinctly advantageous if all the gates of a complex digital function were to be combined in a single package. Resulting systems would be cheaper to make and occupy less PCB space than one which needed many chips to realize the same combinational logic function. Since there would be fewer external components in such systems, reliability would be enhanced. In electronic terms, the shorter interconnections that result in such a high density device would present lower propagation delays and reduce the stray capacitance between adjacent paths.

Fan-in and fan-out. These are measures of a gates input and output loadings. Fan-in is the number of inputs a gate can handle, while fan-out is a measure of a devices output performance, typically the number of other gates which can be connected whilst maintaining specified output levels. Ideally, both values should be high.

Noise immunity or noise margin. This is the maximum amplitude of electrical noise that does not cause an erroneous output when combined with an input signal. Electrical noise may be due to strip lighting, power supplies, current loading from other devices, rotating machinery and atmospherics, and may be minimized by the use of decoupling capacitors across the supply pins of an IC and by ensuring that unused inputs are tied either to ground or V_{CC}. In extreme cases precautions against long tracks behaving as aerials may be required.

In addition to these five properties, other desirable features are:

- circuits should be small and light;
- spares should be readily available;
- system design using standard parts should be easy;
- power supply voltages should not be critical;
- range of devices should be well documented;
- devices should be available from several manufacturers;
- performance should be consistent and reliable over long periods;
- circuits should be physically robust and not easily damaged in normal use or manufacture; and
- devices should be capable of operating in hostile environments.

Technologies

Now we consider a variety of logic circuit families and see to what extent they meet the stated requirements.

Early forms of logic circuits were DCTL (direct coupled transistor logic), RTL (resistor–transistor logic) and DTL (diode–transistor logic). These formats are now obsolete, principally because of their slow operating speeds, high power requirements, poor noise immunity and very low packing densities. The two most popular logic families, currently in use are TTL (transistor–transistor logic) and CMOS (complementary metal oxide semiconductor). We shall concentrate on these two.

TTL (transistor–transistor logic)

This was a popular and widely used logic family. It combined fast speed with moderate levels of power consumption and reasonable levels of noise margin. It is now obsolete, having been replaced by several enhanced variations. The ranges of 74xx based devices are now standardized around specifications originally introduced by Texas Instruments which may be recognized by the '74' prefix to the serial number.

Some principal variations are:

- 74LSxx – low power Schottky;
- 74Hxx – high speed TTL (now being phased out); and
- 74Nxx – standard TTL (sometimes without the 'N').

The essential differences between them are in the component values that are used in the circuits; the features of TTL being broadly the same in each case, but with trade-offs that affect other parameters.

Other variations are:

- 74ALSxx – advanced low power Schottky, which has taken over from 74Hxx;
- 74ABTxx – advanced hybrid, with TTL inputs and CMOS outputs; and
- 54xx – TTL equivalent, but to military specifications (now obsoleste).

TTL noise margins. These are quoted at 1 V (but worst case conditions may be 0.4 V), i.e. the input of a TTL inverter, held at logic 0, may be subject to a 1 V transient without affecting the logic level at its output.

TTL operation. This depends upon transistors being able to switch between cut-off and saturation states.

Common emitter output characteristic. This shown in Figure 8.1. The transistor is off when the operating point is at the lower end of the load line (A). This happens when 0 V is applied to the base. The on condition is when the transistor is driven hard into saturation by a large positive voltage applied to the base (B). In this way an ideal transistor may make an excellent switch in that it can be cut-off to within a few megohms and turned on to a minimum resistance of a few ohms.

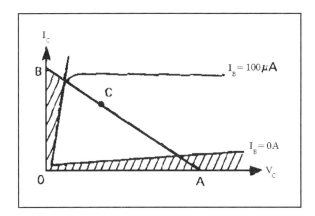

Figure 8.1

TTL NAND gate

A TTL NAND gate is shown in Figure 8.2.

Description:

- T1 is the input driver and is usually a multi-emitter device (only found on integrated circuits);
- T2 is a phase-splitter, having a well defined threshold level; and
- T3 and T4 are the output amplifier pair.

Operation. There are two conditions to consider;

- Both inputs A and B high:
  - T1 collector-base forward biased providing base current for T2;

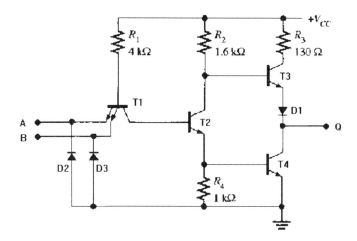

Figure 8.2

- T2 is thus on (in saturation);
- T2 emitter voltage high enough to drive T4 into saturation;
- T2 collector goes low cutting off T3, and therefore output from T4 is low.
- One or more inputs low:
  - T1 operates as normal but drive is insufficient to hold T2 and T4 in conducting state, they are therefore off;
  - T3 turned on as T2 collector voltage rises, and therefore the output from T4 is high.

TTL output configurations

The configuration of T3, D1 and T4 in Figure 8.2 is known as a *totem pole* output amplifier, because T3 and D1 sit on T4. D1 provides a small voltage drop between the T3 emitter and the T4 collector (about 0.7 V), without which T3 would be permanently in saturation. This configuration allows high speed switching and saves on power consumption. It also has good load driving capability (a high fan-out). These outputs cannot be connected together directly, and usually have insufficient fan-out for high current applications.

An *open collector* configuration is used where it is desirable to be able to connect outputs together. Also it is usually designed to be capable of sinking a higher current than the totem pole output. It is obtained by omitting T3 and D1 from the totem pole output, thus increasing the size of T4. The load goes directly between V_{CC} and the T4 collector. When such a gate's output is linked to other gate inputs an external resistor must be connected to V_{CC} in parallel with the load. The value of this resistor depends upon the required fan-out. This output configuration tends to consume more power. It also has poorer noise characteristics and is slower in speed than equivalent conventional gates. However, it is well suited to current sinking applications such as driving common anode displays (LEDs), relay coils or transmission lines. It is used in circuitry that converts signal levels between logic families, as T4 can be designed to withstand a high V_{CC}. The open collector provides for 'wired' logic functions. Here the outputs of four 2AND gates are connected or wired together permanently to give the larger logic function Q = ABCDEFGH.

Wired-AND

A wired-NOR arrangement is derived by replacing the AND with NOR gates (Figure 8.3). Any high input on one of the NOR gates will drive all the outputs low. Wired logic technology is not used much these days, except where logic drives interrupt lines in a microcomputer system.

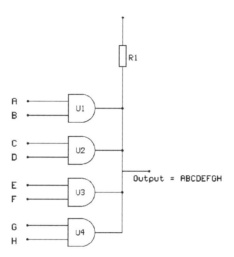

Figure 8.3

Tri-state devices

These amplifiers offer an extra state in addition to the usual logic 0 and logic 1. This extra state is known as a high impedance or tri-state, and is used in situations where more than one gate output is connected to a common line-busses in microcomputers for example. This third state effectively 'switches off' or disconnects the device when it is not required, thus ensuring that its output does not interfere with the current activity. A control or enable line is provided to switch the device on when it is required.

Tri-state NAND gate

A tri-state NAND gate is shown in Figure 8.4. D1 and D2 turn off the output pair of transistors (T3 and T4) if a logic 1 is applied to the enable input, thus giving a high impedance output. This line is, in effect, an active low enable input.

Schottky TTL

Transistors of the Schottky type are used rather than the planar bipolar type. In the Schottky transistor an extra diode of low forward voltage is linked between collector and base. This diode tends to reduce 'carrier storage effects' when the transistor is switched fully on. This results in a higher speed (lower propagation delay) version of TTL.

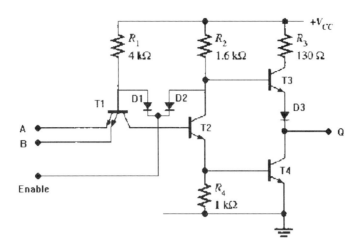

Figure 8.4

CMOS (complementary metal oxide semiconductor)

CMOS combines both NMOS and PMOS technologies where the gates are constructed with mixed N and P channel MOSFETs. They are enhancement FET types in which no conducting channel exists between the source and the drain in the absence of a bias voltage. This second versatile and popular logic family features very low power consumption, a wide operating supply voltage range, a very high fan-out and excellent noise immunity. However, such devices are prone to static electricity damage, and early versions were fairly slow devices. A very low power consumption results in very high packing densities and complex interconnections. For this reason, LSI and VLSI circuits are almost invariably CMOS devices, although a few mix technologies for higher speed.

A large number of CMOS circuits are found in the 4000 and 4500 series. Others are the 4100 and 4700 series. Except where the relatively high operating voltage and tolerance of supply variation is an advantage, 4xxx devices have become obsolete. The replacement series follow the '74' naming convention of TTL, with the addition of improved versions of the original 4xxx devices such as, for example, 74HC4xxx. Available families include:

- 74HCxx – high speed CMOS;
- 74HCTxx – TTL input compatible HCMOS;
- 74BCTxx – TTL input compatible bus oriented HCMOS;
- 74AHCxx – even higher speed CMOS; and
- 74AHCTxx – TTL input compatible AHCMOS.

You may meet:

- 74Cxx – the original CMOS version of 74xx (now obsolete); and
- 74ACxx – a high speed version of 74Cxx (now obsolete).

CMOS noise margin

This is one of the best features of CMOS. It is normally quoted as a percentage of the supply voltage – typically 45%. This means that a CMOS device with a 5 V supply can tolerate a 2.25 V spike. The exception is HCT, which only has TTL immunity as the inputs are designed to be TTL logic level compatible.

The inputs of CMOS logic family gates are connected to the gates of MOSFETs and are insulated from the remainder of the circuit by a very thin layer of SiO$_2$ (silicon dioxide). Breakdown of this layer and consequent destruction of the circuit occurs if a gate potential is allowed to rise above 50 V. Static charges of several thousand volts may build up on objects insulated from ground, and if this were allowed to happen on CMOS inputs, damage would be inevitable. Nowadays, CMOS inputs are protected to a degree against static build-up by use of protection networks (Figure 8.5); however, it is still essential to follow static avoidance practices.

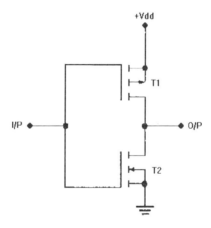

Figure 8.5

CMOS inverter

With a logic 0 applied to the input, T2 (N-channel) will be cut off, since there is zero volts between its gate and source, and T1 (P-channel) will be on, because there is a large negative bias between its gate and source. The on resistance of T1 is about 300 Ω, while the off resistance of T2 will be an extremely high (in excess of 100 MΩ) - virtually open circuit. With a logic 1 applied to the input, the reverse happens.

CMOS NAND and NOR gates

These are simple extensions of the inverter described above. For the NAND gate (Figure 8.6), the output can only be low when both T3 and T4 conduct and both T1 and T2 are off. This condition only occurs when both inputs A and B are at logic 1. For

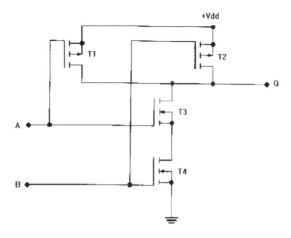

Figure 8.6

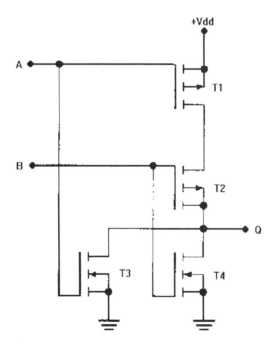

Figure 8.7

the NOR gate (Figure 8.7), the output will be low if either or both the inputs A and B are a logic 1.

A 200 Ω resistor and two diodes (not shown) form a built-in protection circuit; these have been omitted from the NOR and NAND gates for clarity. All unused inputs must be connected somewhere (either to V_{DD} or ground) – they must not be left floating.

Inputs		State of MOSFETs				Output	Output
A	B	T1	T2	T3	T4	NAND	NOR
0	0	On	On	Off	Off	1	1
0	1	On	Off	Off	On	1	0
1	0	Off	On	On	Off	1	0
1	1	Off	Off	On	On	0	0

Other logic families

ECL (emitter coupled logic)

Like TTL, ECL is constructed using bipolar transistors. In TTL circuits a transistor is driven from cut-off to saturation to represent logic 1 and logic 0, respectively. This limits the speed of the device. ECL achieves higher speeds by driving transistors between a point in the linear operating region and cut-off, but at the expense of very much higher power dissipation. ECL is usually confined to the larger installations where heavier duty power supplies are available and where no expense is spared. Output levels are typically $-0.8\,V$ for logic 1 and $-1.8\,V$ for logic 0. As improved manufacturing techniques allow for faster CMOS devices, this logic is gradually disappearing, except for radiofrequency applications, where similar devices based on gallium arsenide (GaAs), rather than silicon, are gradually appearing.

IIL (I^2L; integrated injection logic)

This is widely used in pocket calculators where VLSI is required. It is an advanced version DCTL which uses bipolar transistor logic and rivals the low power and high packing density of CMOS.

Which family to use?

This really depends upon the principal needs of your system, especially whether you want speed or just low power consumption. If speed is not an important criterion, then CMOS is the best bet, with a comprehensive range of devices. It is possible to mix devices of several families in one system or circuit, but attention must be paid to interfacing the different signal and supply voltage levels. In general, CMOS devices may be used in a predominantly TTL circuit without too much difficulty, and since CMOS will generate TTL signal levels given a 5 V supply, interfacing does not present a problem. As relatively few devices in any system operate at maximum speed, it is always wise to choose devices with a low static dissipation where power or heat is likely to be a problem.

Today, discrete logic chips are not used in complex digital systems. Programmable devices such as the PAL, PLD, FPGA and ASIC are now commonplace. The

Table 8.1

	Low power TTL	High speed TTL	CMOS	HCMOS	DTL	Standard TTL	ECL	I²L
Typical device	74LS00	74H00	CD4011	74HC00	Obsolete	7400	MC1662	Special
Speed*	30 MHz	50 MHz	1 MHz	30 MHz	30 MHz	35 MHz	1000 MHz	1 MHz
Propagation delay	10 ns	6 ns	15 ns	10 ns	30 ns	10 ns	0.5 ns	250 ns
Power dissipation*	2 mW	22 mW	10 nW	10 µW	11 mW	10 mW	60 mW	70 µW to 6 nW
Power/speed product†	20 pJ	132 pJ	150 aJ	200 aJ	275 pJ	100 pJ	30 pJ	25 pJ
Range of devices*	Medium	Medium	Large	Very large	Small	Medium	Small	Small
Integration density*	Medium	Medium	High	High	Low	Medium	Low	High
Fan-out/fan-in*	20	10	50	>>100	8/14	10	25	n/a
Output impedance	Low	Low	High	Medium	High	Low	Low	
Noise immunity	800 mV	800 mV	45% of V_{CC}	45% of V_{CC}	1.2 V	1 V	500 mV	350 mV @ 1 V V_{CC}
Operational chars	Saturating bipolar transistor	Saturating bipolar transistor	Saturating MOSFET	Saturating MOSFET	Diode and saturating transistor	Saturating bipolar transistor	Non-saturating bipolar transistor	Saturating bipolar transistor
Basic gate	NAND	NAND	NOR/NAND	NOR/NAND	–	NAND	NOR/OR	–
Static problems*	None	None	Some	Some	None	None	None	None
Logic 1	> +3.3 V	> +3.3 V	V_{CC} to +0.7	75–120% V_{CC}	+2.3 > 6.3 V	> +3.3 V	−0.74 V	Approx. V_{CC}
Logic 0	< +0.2 V	< +0.2 V	0 to +0.3 V	−20% to 25% V_{CC}	0 > 0.8 V	< +0.2 V	−1.6 V	0.4 V
Power supply*	+5 V @ 5%	+5 V @ 5%	+3 to +18 V	+2 to +6 V	+6 V @ 5%	+5 V @ 5%	−5.2 V	+1 to +15 V

The list is not comprehensive, as manufacturers are developing new variants of logic families all the time. There is a trend towards 3.3 V supply systems, as this reduces internal isolation required, allowing smaller geometries and yielding increased speed with a given technology.

* Principal considerations.

† 1 atto joule (aJ) = 1×10^{-18} Joule (J).

ULA (uncommitted logic array), PLA or PAL (programmable logic array), PLD (programmable logic device) and FPGA (field-programmable gate array) are VLSI chips containing many thousands of gates the interconnections of which are not fixed during manufacture but may be set by the user in a special programming sequence. Other device types encountered today are ASICs, (application specific integrated circuits). The user will use a programming language such as ABEL, PALASM or VHDL to derive the necessary logic combination. Whereas PALs are programmed by the user in-house, ASIC designs have to be synthesized onto silicon by specialists. A few types can be dynamically reconfigured rather than being single-use devices.

The characteristics of the various logic families are compared in Table 8.1.

Handling integrated circuits

Care must be exercised when handling ICs – especially CMOS devices. Static damage may not be catastrophic, or immediately apparent. Device performance is easily degraded, and unpredictable early device failure may be caused.

- When handling an IC, be sure you are connected to the system ground potential – ideally by using proper static safe equipment.
- Always store CMOS ICs with their pins shorted together by embedding them in conductive foam or aluminum foil, or in a special IC holder.
- Do not remove the IC from protective storage until it is required for insertion into the circuit.
- Do not touch the pins of a CMOS IC.
- Do not remove or install a CMOS IC from a circuit whilst it is powered.
- Use a grounded tip soldering iron when installing the IC – better still, use IC sockets for prototyping. (*Note*: IC sockets are rarely used in production designs because of reduced reliability and increased cost.)
- Do not apply input signals while the power is disconnected.
- Do not leave unused CMOS IC inputs floating.
- Avoid use of materials that encourage the build-up of static electricity.

Practical Exercise: creating integrated circuit modules

(*Note*: This section requires a full version of EASY-PC Pro, or Pro XM. The supplied trial version will not allow component creation, though you may create the initial symbols. The EASY-PC Pro menu structure differs from XM, so several commands will vary if you are using this.)

As you develop your use of EASY-PC Pro and PULSAR you will want to generate IC libraries of your own and use these modules in circuit designs. You have already had a taste of this topic in Chapter 5 when you produced a D-type flip-flop called DT7474. Reference to a TTL data book will confirm that the circuit you drew was one-half of a SN74LS74. Here we will complete the full TTL component.

Several other TTL devices will be suggested also. However, it is worth noting that Number One Systems are able to supply very full libraries of TTL, CMOS and surface mounted devices at a reasonable cost, so there is no need to tackle all the devices in the data book.

This section will show you how to create a PCB component from basic principles. You will need to follow the steps described in Chapter 4, which cover the creation of a schematic symbol and component, and then the simulation of results. You are advised to save any modules you create to your own libraries.

Logic circuit ideas

- SN7493 — 4-bit binary counter (divide by 2 and 8).
- SN7476 — dual JK flip-flop.
- SN7474 — dual D-type flip-flop.
- SN74138 — three- to eight-line decoder/multiplexer.
- SN74194 — 4-bit bi-directional universal shift register.

4-bit binary counter (SN7493)

Follow the instructions in Chapter 4 and draw the circuit shown in Figure 8.8. Use the negative edge-triggered JK (called NJK) provided in the PULSAR library. Remember to avoid '/' in net names. You must tie all NP and JK inputs HIGH (to V_{CC}).

Now you must simulate the circuit in PULSAR. A low signal on both R0 lines will enable the circuit. Section A is a divide-by-2 counter so the output at QA will be half that on input A. Check this by applying a 10 kHz generator to input A. Section B can be checked by linking the output at QA to input B. This will deliver a divide-by-16 count at QD.

When you are satisfied with the circuit's performance, create a PULSAR module called SN7493 and save it in your USER.PLB library. Give the NJK elements a propagation delay of 16 ns and the 2NAND a delay of 10 ns. The procedure for all this is described in steps 36–43 in Chapter 4.

Create a schematic symbol and schematic component for SN7493 following the procedure described in steps 44 onwards in Chapter 4. Figure 8.9 shows a suggested schematic symbol. Name it SN7493 and give it the reference type of 'IC'. When you are satisfied with the result save it to USR_IC.SIC. Figure 8.10 shows a suggested schematic component. Name it SN7493 (as before) and describe it as a Binary Counter IC. The package type will be DIL. Save this to your own USR_IC library. When you have completed this, test the component by following steps 85–94 in Chapter 4.

Continued on p. 186

Practical Exercise: creating integrated circuit modules
(*Continued*)

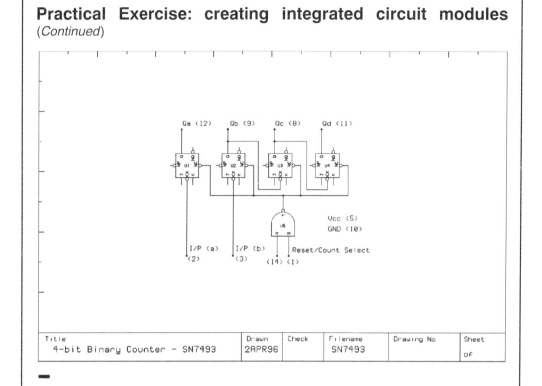

Title		Drawn	Check	Filename	Drawing No	Sheet
4-bit Binary Counter - SN7493		2APR96		SN7493		of

Figure 8.8

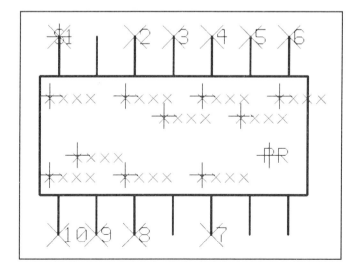

Figure 8.9

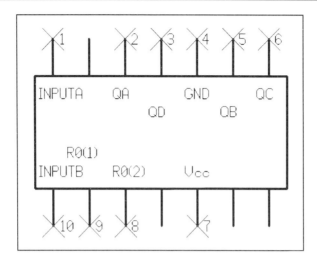

Figure 8.10

Creating a PCB component

Step 1: Return to the Mode Selection menu: `SHIFT` `F10` `Y`.
Step 2: The PCB symbol has already been defined for you—14DIL—so hit `D` to begin PCB component creation (Figure 8.11).

```
        Mode Selection
 Schematic                 A
 PCB                       B
 Sch Component             C
 PCB Component             D

 Sch Symbol                E
 PCB Symbol                F

 Exit To DOS               X
```

Figure 8.11

Step 3: Select New component (Figure 8.12).
Step 4: Select the Component menu (Figure 8.13).
Step 5: Work your way down the list, naming it SN7493 and describing it as a Binary Counter IC as before.

Continued on p. 188

Practical Exercise: creating integrated circuit modules
(*Continued*)

```
┌─────────────────────────────┐
│  Component Operations       │
│  New                      N │
│  Edit                     E │
│  Browse                   B │
│ ─────────────────────────── │
│  Delete                   D │
│ ─────────────────────────── │
│  Press ESC to quit          │
└─────────────────────────────┘
```

Figure 8.12

```
┌──────────────────────────────────┐
│          PCB Component            │
│  .  ┆ Name              ┆         │
│  .  ┆ Description        ┆         │
│  .  ┆ Load Symbol        ┆         │
│  .  ┆ Part Number        ┆         │
│  .  ┆ Package Type       ┆         │
│ ──────────────────────────────── │
│  .  ┆ Pin(s)             ┆         │
│ ──────────────────────────────── │
│  .  ┆ Add to Library     ┆         │
│ ──────────────────────────────── │
│         Quit              OK       │
└──────────────────────────────────┘
```

Figure 8.13

Step 6: Load symbol 14DIL from the supplied library called IC.PIC.

Step 7: The package type will, of course, be DIL.

Step 8: When you come to name the pins, it is worth centering the outline first and work at zoom level 3 or 2. Remember you [ESC] from this menu and then use the Pins menu.

Step 9: Name each pin as shown in Figure 8.14 by clicking on each pad in turn and entering text in the panel provided. Define Gates and Define Equates are used when there is more than one identical element on a chip, and so are not needed here.

Step 10: When this is completed, re-enter the PCB Component menu (Figure 8.15).

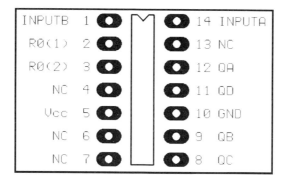

Figure 8.14

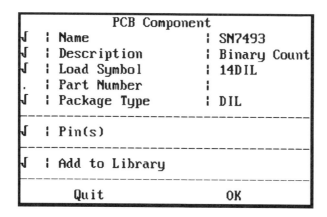

Figure 8.15

Step 11: Check that each operation has been ticked off in the left-hand column.

Step 12: Add your new component to your USER library then return to the Mode Selection menu by typing `Q`, `Y`, `ESC`.

Step 13: You can check this module by loading the SN7493 into a schematic. Save this circuit as SNTEST.SCH for use later on, and then translate it into a PCB.

Dual D-type flip-flop (SN7474)

In the 14-pin DIL device shown in Figure 8.16 there are two D-type flip-flops. The numbers in the brackets refer to the DIL pins for each section: section 1 is given first with the supply pins shown in the top right-hand corner. Give the inverting output the name NQ.

Continued on p. 190

Practical Exercise: creating integrated circuit modules
(Continued)

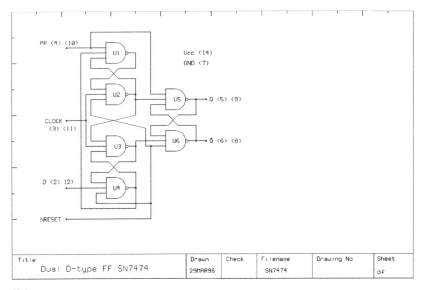

Title		Drawn	Check	Filename	Drawing No	Sheet
Dual D-type FF SN7474		29MAR96		SN7474		of

Figure 8.16

The SN7474 is a gated device, so you will need to include a positive-edge pulse shaping circuit on the clock input line. Refer back to Figure 5.11 to see how you did this.

Dual JK flip-flop (SN7476)

Like the D-type flip-flop described above, this is a gated device (Figure 8.17). This time you will have to include a negative pulse shaping circuit. You can test it the same way as the JK flip-flop described in Chapter 5. Again, use the prefix 'N' to indicate inversion, e.g. NCLK.

This JK flip-flop is also a dual device, but this time with 16 pins. Use the 16DIL PCB symbol.

Three- to eight-line decoder/multiplexer (SN74138)

This circuit (Figure 8.18) will decode one of eight lines (Y0 to Y7) depending on the condition on the three binary select inputs (A, B and C). Three enable lines are provided, one active high and the other two active low (G1, NG2 and NG3). These allow expansion up to larger decoder arrays without the need for additional external select logic.

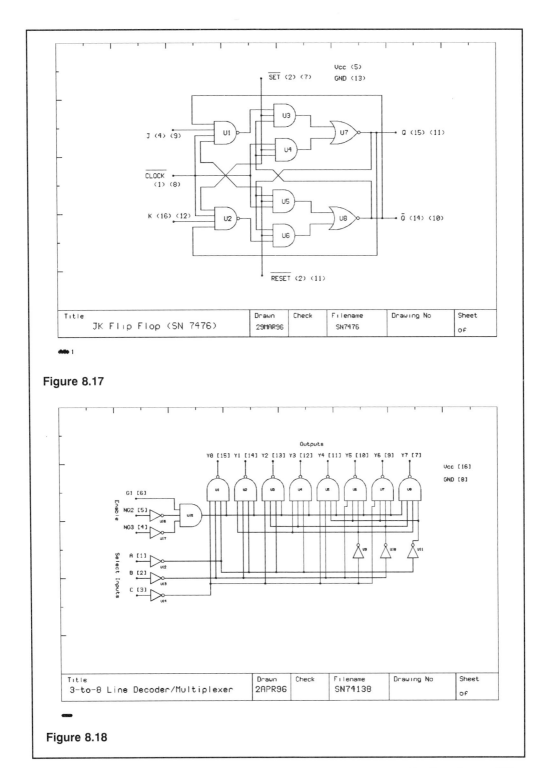

Figure 8.17

Figure 8.18

Continued on p. 192

Practical Exercise: creating integrated circuit modules
(*Continued*)

To simulate the device apply 2-, 4- and 8 kHz generators to inputs A, B and C, respectively. Tie G1 high and NG2, NG3 low to turn the device on. PULSAR should deliver the waveform shown in Figure 8.19. Reverse the generator signals on A and C if the output signal progression appears to be incorrect

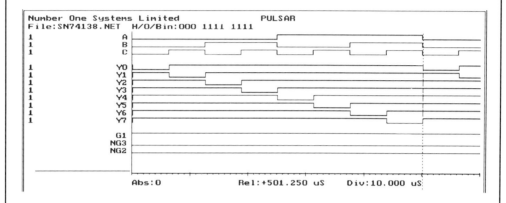

Figure 8.19

9 Printed circuit boards

Introduction

All the components in a digital electronic system, including the integrated circuits (ICS), are usually connected together on a printed circuit board (PCB) of some kind. As the performance and complexity of these digital systems have increased, the design and variety of PCBs have had to evolve to keep pace with the latest requirements for miniaturization and speed. Nowadays the PCB has almost as much relevance on the performance of a digital electronic system as the discrete components used on it. Currently, PCBs range from the single sided type, with components mounted on the top side and conductor tracks on the bottom, to multi-layer types which have conductor layers laminated within the board, very similar to the layers in plywood.

Printed circuit board design

Component placing and track routing used to be a manual process which was very slow and painstaking, requiring a great deal of skill. Artwork was produced by sticking crepe tape outlines on thick (0.005–0.007 inch) Melinex-based draughting film (sometimes known as Permatrace). The resulting image was then photographically transferred onto the copper laminate of a PCB.

With the advent of personal computers, artwork was soon being produced using computer draughting packages. These packages were not only much quicker, but usually more accurate, and more easily checked and changed when modifications were necessary. Nowadays most complex PCB layout design is done by computer aided design (CAD) processes; however, a few companies still prefer to route their circuits by hand. (The principal disadvantages of CAD are the inability to view the whole design area simultaneously at high resolution, the restriction that pads cannot be trimmed where clearances are tight, the need to generate new library parts where unusual components are needed, and the relative lack of portability of the necessary computing equipment.)

The artwork is normally laid out in a standard grid pattern of 0.1 inch. This matches the lead or connector pitch used by most components, in particular the standard integrated circuit. For surface-mount technology, where component leads are not taken through the PCB, the grid is considerably smaller (1 mm (0.04 inch) or 0.025 inch). As a rough guide, non-standard footprints of American or British origin tend to be Imperial, while continental and Far-Eastern sources normally use metric grids.

Each component has an outline of connections or pads that represent the position of its leads. These pads exist in a variety of shapes and sizes. They represent the area on the PCB that will be used to make a solder connection to the copper track, and usually have a marker that indicates the position of a hole which will be drilled or punched through the board during the production phase.

Most manual designers use artwork which is twice or four times full size, with pads, tracks and other artwork scaled up accordingly. This allows a greater degree of accuracy, but requires photographic reduction before it can be used to produce the final circuit board. For simple applications most artwork is designed at full size so that it can be transferred to the final circuit board very easily. The zoom ability of CAD packages usually renders this unnecessary.

Many computer PCB design packages exist, ranging in price from about £100 up to many thousands. Favoured methodology allows for the circuit diagram to be drawn first. Then a 'schematic capture' will produce a net-list. This itemizes all components and interconnections and contains the sizes, types or ratings of the components to be used in the circuit, and may even contain the outline of the PCB itself.

This net-list will then be translated into PCB form, where all components and connections appear like a 'rats nest' in the corner of the drawing board. The designer can then place the critical components such as edge-connectors, ICs, etc., at their preferred places on the PCB, and then set about arranging the remainder and routing the conductors as best he or she can. Ultimately, the designer could opt for an autoroute process where the computer makes the best guess at routing virtually all tracks on the circuit board. This may speed up the design of a PCB layout, but is heavily dependent upon design rules, component placement, board density, and (depending upon the type of package used) can take some time. However, it is much faster than manual routing, and most allow trial passes so component placement can be improved without a lot of manual rework. Depending on the factors mentioned earlier, it is not unusual for autorouters to fail to route a board fully. If this does happen, it is normally the most difficult routes that remain, so some ingenuity may be required to complete the board manually. However, as we shall see, the tracks themselves play a vital part in the performance of the circuit, and autorouting is not always the best answer.

EASY-PC Professional represents the best value for money at the lower end of the PCB CAD market. It follows the popular methodology just described, although the budget or entry-level package does not have an 'autorouting' function. The more advanced version, EASY-PC Professional XM, does support an optional autorouter, called MultiRouter. This book refers to EASY-PC Professional XM throughout, as the disk supplied includes a trial version of this program. If you are using a full version of EASY-PC Professional, you will already have noticed several differences between the menus and hot keys in your software and the text.

Types of printed circuit boards

We will consider five of the most common types of circuit board: single sided boards, double sided boards, double sided boards with plated through holes, multi-layer boards and flexible boards.

Single sided boards

These are the simplest types and comprise a thin copper laminate (roughly 35 μm) bonded to a saturated (or synthetic) resin bonded paper (SRBP) or epoxy–glass board roughly 1.5 mm thick. Single sided boards, especially the SRBP type, are used in the low-cost domestic electronics industry and, as their name suggests, the components are mounted on one side and soldered to copper connections on the other. They are particularly suited to project or non-professional prototyping work where circuits are not too complex, but they have the disadvantage that the more complex circuits may require wire 'jumpers' to make all the desired connections. These boards present a greater challenge to autorouters because of the difficulty of finding routes that do not cross others. Figure 9.1 shows an autorouted single sided board.

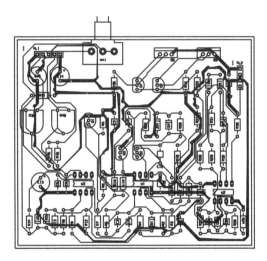

Figure 9.1

Double sided boards

These are made from the same material as single sided boards and have copper laminates on both sides. Components are usually mounted on one side with their leads providing electrical connections to additional tracks on the top surface. Double sided boards are ideal for the more complex prototype boards and where jumper connections are unavoidable. The main problem encountered is lining up both top and bottom tracks which may be quite difficult at the workshop level. Figure 9.2 shows the same circuit as in Figure 9.1, but in a more compact double sided board. Note how all the through board connections can be made by soldering components on both sides.

Double sided boards with plated through holes

These boards (Figure 9.3) are very common in industrial electronics, particularly in calculator and computer circuits. They permit very high component densities,

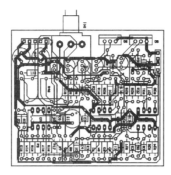

Figure 9.2

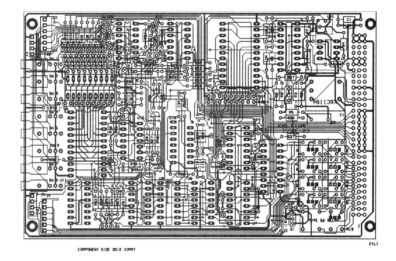

Figure 9.3

with every hole providing a conduction path between the top and bottom surfaces. Additional special holes, called *vias*, not used by component leads, provide connections between the two sides. These have smaller pads and holes too narrow for components. Boards with plated through holes tend to be more expensive than the standard double sided types, as every additional hole adds to the cost. This is the preferred technology for most autorouters, as one surface can be tracked predominantly in the x axis, with the other predominantly in the y axis. Vias are used to change direction.

Multi layer boards

The growing demand for greater component densities on smaller PCB outlines soon makes the double sided PCB solution very difficult, if not impractical. The designer has now to face many unpredictable design problems such as noise, stray capacitance and cross-talk. A multi-layer board offers a suitable solution. Four, six and ten layers are common, while there are even some types in excess of 30 layers, making them somewhat thicker than the conventional single or double sided board. Electrical connections between the different layers are achieved with vias or plated through holes. Where the inside connections are not required, the diameter of the roundel on that plane is increased and the hole drilling or punching process will clear enough copper to prevent connection.

The simplest type is a four-layer board where the middle two layers are reserved for the supply planes with the other two being for component connections. This arrangement uses the earth and power planes as a screen or shield for each signal conductor and thus minimizes cross-talk. It also provides a high distributed capacitance which helps to decouple the power rails. A frequent application for this technology is in high speed computer motherboards.

Construction is by sandwiching the appropriate number of (very thin) double sided boards together with insulating separators. These are epoxy glued, very accurately, using heat and high pressures. Vias can be restricted to one double sided element (blind and buried vias), or by a post-plating process may pass through the board, connecting to selected layers. It follows from this that multi-layer boards usually have an even number of layers.

Multi-layer boards represent roughly a third of the PCB market and are naturally very expensive. However, their application can be cost effective, since much greater packing densities and improved frequency response over double sided boards are now possible. See the detail in Figure 9.4.

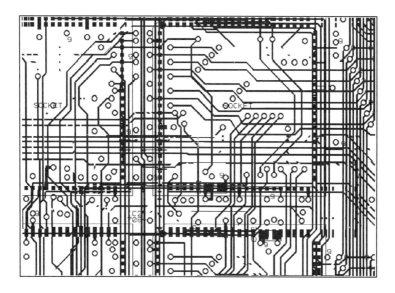

Figure 9.4

Flexible boards

Flexible boards account for about 10% of the PCB market, their flexibility and weight giving them an economic advantage over the rigid types. They are usually used as connection harnesses between several boards in an electronic system, but may well carry components in their own right. Their only real disadvantage lies in their limited ability to handle high frequencies. They are commonly encountered in the flexible loom connecting to dot matrix and ink jet print heads. They are also increasingly appearing in vehicles in place of discrete wire looms.

Flexible boards can be manufactured in either single, double or multi-layer configurations using a range of base materials and conductors, including polyester, polyamide and Teflon foils and fibreglass bases. Each type offers slightly different characteristics such as flexibility and tear resistance.

Solder pad shapes and positioning now become important design considerations. To minimize the risk of conductor breakage it is best to avoid placing solder connections in areas where bending will occur; this also allows as much bending radius as possible. The final product sometimes has added strengtheners or supports and is usually coated with a protective film to increase its robustness (Figure 9.5).

Figure 9.5

PCB manufacture

Once the layout artwork is completed a full size positive transparency is produced. A copper laminated board is prepared with a coating of photosensitive etch-resist emulsion. Artwork is then placed over the emulsified area and this arrangement is exposed to ultra-violet light for a few minutes using a special light-box.

The positive transparency masks the required tracks, while the ultra-violet light chemically changes the exposed areas of emulsion. This altered emulsion is washed away in a diluted alkaline solution and thus fixes the layout image transferred from the artwork. For very high volumes the etch resistant pattern may be silk screened onto the copper. Now the board is placed in a bath of heated acidic ferric chloride solution and the unwanted copper is slowly etched away. (Manufacturing equipment often sprays etchant at the board so that waste residues drain away.) This etching takes a few minutes, the exact time depending upon the strength and temperature of the solution. Once etched, the board is washed, dried, cropped to size and finally drilled.

Take care not to overexpose the board to ultra-violet light in the first instance or else 'fringing' will occur and cause adjacent tracks to short together. Timing is also vital when the board is being etched in the ferric chloride bath. Overetching will cause depleted track widths at the very least, and will completely clear the copper laminate if left in the bath too long.

Professional circuit designers may wish to coat each board surface with a solder resist compound. Solder resist is a coloured lacquer (usually green or blue) that covers the entire circuit board, only revealing areas where solder connections are to be made. It prevents solder bridges, especially where flow-solder techniques are used. The solder resist mask entails an additional drawing for each copper surface, which just shows the areas where soldered connections are to be made.

For prototype or one-off boards hole drilling is normally done by hand. Where large volumes are needed a manufacturer will use automatic drilling machines at this stage or even a punch tool for extremely large quantities in low tolerance, price sensitive applications.

Prototype double sided boards are produced in much the same way except that care must be taken to register both the top and bottom artwork correctly when exposing the emulsified board to ultra-violet light. Extra location holes beyond the PCB outline are drilled to ensure that the two images line up. Care must also be taken to mask the opposite side when the other is being exposed to the ultra-violet light.

Double sided boards with plated through holes require a rather different process. The most widely used method of manufacture is the 'pattern-plating' process. Here the holes are drilled first, and then a very thin layer of copper is deposited over the entire surface of the board, including all the holes. The board is coated with a photosensitive emulsion and then exposed to ultra-violet light, with the through hole areas masked by positive artwork. The unexposed emulsion is removed, revealing the copper areas of the pads and vias. Another electrolytic process thickens the copper areas to roughly 35 μm, after which the board is tin-plated to allow better solderability. A final etching process clears all unwanted copper off the board leaving the tin-plating on the tracks, pads and through holes.

(*Note*: The terms 1 ounce copper, 2 ounce copper, which you may still encounter, refer to the weight of copper covering the board before etching. 1 ounce copper is equivalent to a 35 μm coating. Other values are pro rata.)

PCB parameters

Today's PCBs are so complex that they has to be viewed as a separate electronic components with their own circuit characteristics. Simple low frequency applications pose few problems for the PCB designer, but at higher frequencies, especially those used in digital systems, these parameters become very significant.

The parameters of a PCB are quite likely to affect the responses of very high frequency pulses, and the high surge currents which accompany them can court disaster if thin copper tracks are used. The tracks themselves present a certain line impedance, the capacitance of which can cause cross-talk to occur between adjacent signal paths, and the inductance of which can cause pulse distortion. Together, impedance mismatching can occur, giving rise to standing waves and ringing.

The PCB then is a complex component in its own right, comprising elements of resistance, capacitance and inductance. The transmission line effects that these elements cause are of great concern to the designer. We will now investigate these three properties and then see how they relate to transmission lines.

If copies of ANALYSER III Professional and LAYAN are available, it can be very instructive to simulate an area of a PCB with and without taking the effect of the board into account. Especially as frequency increases, the difference can be quite marked.

Resistance

The resistivity of copper is given as $1.7241 \, \mu\Omega \, \text{cm}$ at $20°C$ and is sufficient to cause a significant voltage drop along the length of a PCB track. The resistance of the track can be calculated quite easily as:

$$\text{Resistance } (\Omega) = \frac{\text{Resistivity of track } (\Omega \, \text{m}) \times \text{Length of track (m)}}{\text{Cross-sectional area (m}^2)}$$

Assuming that the thickness of the copper laminate is $35 \, \mu\text{m}$ (a typical figure), the resistance of a 1 cm long track, 1 mm wide will be roughly $5 \, \text{m}\Omega$. More precisely:

$$R = \frac{r \times l \, (\Omega)}{A} \quad \text{or} = \frac{1.7241 \times 10^{-8} \times 0.01}{35 \times 10^{-6} \times 0.001} = 4.9 \, \text{m}\Omega$$

A useful rule of thumb is that a 20 cm long track has a resistance of about $100 \, \text{m}\Omega$. When the track width is increased the resistance decreases. If the track width is 1 cm instead of 1 mm, the resistance is $490 \, \mu\Omega$.

Capacitance

Where two tracks are placed close to one another on a PCB they will exhibit a capacitance effect in exactly the same way as two conducting plates will do in a capacitor. This applies to adjacent conductors on the same side of the board as well as those on opposite sides, where the PCB material provides a useful dielectric or insulating medium.

Calculating the capacitance between tracks on opposite sides of a PCB is very straightforward. We just use the general equation for capacitance:

$$\text{Capacitance (farad)} = \frac{\text{Absolute permittivity} \times \text{Overlapping track area}}{\text{Board thickness}}$$

$$C = \frac{\varepsilon_0 \varepsilon_{\text{r}} A}{d}$$

where ε_0 is the permittivity of free space, $(= 8.854 \times 10^{-12} \, \text{F/m})$ and ε_{r} is the relative permittivity of the dielectric medium.

The relative permittivity of a PCB is somewhere between 4 and 8, depending upon the material used. For two 3 mm wide tracks, 10 cm in length, on opposite sides of a PCB (nominally 1.6 mm thick) the capacitance will be roughly 10 pF. This capacitance will be larger as the board thickness decreases or as the cross-sectional area of the tracks increases.

The mathematics needed to calculate the capacitance between adjacent conductors is rather complex so we must take a practical approach. It is about 0.4 pF per centimetre length of track. The actual value will depend upon the thickness of the copper used and the widths of the tracks. Separation only becomes a significant factor once the tracks are closer than 1 mm.

With two $35 \, \mu\text{m}$ thick conductors, each 1 mm wide and separated by a 1 mm gap, the capacitance over a 15 cm track length would be roughly 6 pF. Capacitances of this order can be both a help and a hindrance. The 10 pF may serve as a convenient decoupling between supply rails and ground, but can induce unwelcome interference between signal tracks.

Inductance

Here again the mathematics required to calculate inductance is too advanced for this course. The actual value depends upon the conductor width and spacing, with the relationship being directly proportional to spacing and inversely proportional to width. In other words, inductance increases as the spacing increases, and decreases as conductor width increases.

Two 1 mm wide conductors separated by 1 mm would have an approximate inductance of 8 nH (nanohenrys) per centimetre. Doubling the spacing will increase the inductance by 2 nH, as will halving the track width.

Transmission line

The above three parameters – resistance, capacitance and inductance – appear together throughout the lengths of all tracks on a PCB. Their overall effect is to produce miniature transmission lines throughout the circuit and, unless the layout is carefully designed, will seriously affect its performance.

The equivalent circuit for a transmission line section is shown in Figure 9.6. An inductor, L1 has a series resistance R2, and capacitor C1 has a shunt resistance R1 across its dielectric medium. This is complicated further, with the inductor exhibiting capacitance between its 'coils' and the capacitor experiencing mutual inductance between its 'plates'. The overall effect can be simplified to the above circuit. A line comprises a collection of such sections connected in series. On a PCB we are concerned with the effect that the inductance and capacitance properties have in creating transmission lines from parallel tracks.

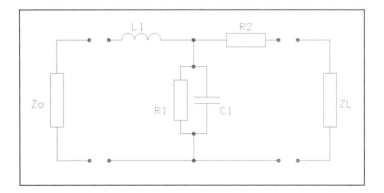

Figure 9.6

Transmission line effects

Mismatching: ringing and reflections. Figure 9.6 also shows Z_0, which is the characteristic impedance of the line, and Z_L, the load impedance. On a PCB the load impedance is what is sensed as the shunting value on each track. So, for a line to be matched, the characteristic impedance must equal the load impedance. The actual value of the characteristic impedance Z_0 can be calculated using the formula:

$$Z_0 = \sqrt{\frac{L}{C}}\ \Omega$$

In practical circuits this value could be anywhere between 50 and 500 Ω. Typically, two 1 mm wide parallel tracks separated by 1 mm would have a Z_0 of roughly 150 Ω.

Mismatching occurs when Z_0 and Z_L are not equal to each other. Now, part of the information travelling along the PCB tracks is reflected back when it reaches its destination, the amount of which depends very much on the degree of mismatch. The reflected signal travels back down the lines to the source, then part of it is reflected back to the destination again; the process is repeated until the reflected signal amplitude eventually dies away.

The circuit designer must be aware that these reflected signals interfere with those following on behind, modifying them, in some cases disastrously. By careful circuit layout the designer can aim for a characteristic impedance somewhere near 150 Ω. This value ensures that at least the integrity of the leading and trailing edges of the transmitted data signals is maintained, even though there is some overshoot. With this information it is possible to restore the shape of the original signal. Unfortunately, overshoot presents us with additional problems. If the characteristic impedance is too low the overshoot at the receiving end increases, especially on the trailing edge, giving rise to ringing.

To overcome reflections designers may use as thin a signal line as manufacturing permits. Matching resistors or line drivers will compensate if the problems are severe. Digital components also present difficulties, as input and output impedances usually vary with frequency, and even the applied voltage. Perhaps the most important point is that high frequency lines should be kept as short as possible. All the above effects are minimized if the track length is short compared to the wavelength of the signal.

Cross-talk. Cross-talk is where two different signals running along adjacent signal paths interfere with each other. This effect can be noticeable on a PCB where two long signal tracks run parallel so one of the conductors can induce a spurious signal on the other.

Cross-talk is not such an issue with CMOS circuits, where input impedances tend to be very high. However, for TTL circuits this effect starts to be a problem once parallel tracks exceed 30 cm. Analysing the effect requires a complex equivalent circuit comprising two or more sets of transmission lines, and presents more of a mathematical challenge than the mismatching effects.

For practical purposes the further apart you can place the signal lines the less the cross-talk will be. It also helps to site them close to ground lines or ground planes to provide suitable screening. The latter course is eminently suitable for double sided or multi-layered boards. With single sided boards, placing a ground conductor between the two signal lines is the best solution. Here the capacitance along the length of this ground line will effectively decouple the signal.

Designing a printed circuit board

Preliminary considerations

Now that the physics of a PCBs has been discussed, we need to explore exactly how to make a prototype circuit board. Assuming that the electronics have been designed, the schematic diagram has been drawn and simulated, the next sensible step is to review exactly what sort of product you are going to make. Firstly, it is important to examine the components that are going to be used. You will need to decide upon those that are going to be mounted on the circuit board and those that are to be fixed to the box or enclosure. Where you mount them is generally a matter of common engineering sense, but by collecting samples of every component or component type it is possible to identify those that will cause layout problems. This is where a parts-list comes in handy. It should contain the details of all the components, their values, ratings and quantities, and reference or catalogue numbers.

If a component is bulky it is best left off the board. For example, low frequency transformers and large electrolytic capacitors should be mounted in the enclosure or on the chassis. The components that generate most heat should also be mounted off the board where possible, while others, such as power transistors, may have to be attached to heat sinks. These off-board components, and items such as the product enclosure itself, switches and controls, and plugs and sockets, will have a direct bearing on the shape and size of the circuit board. You should also consider where and how the board will be fixed into place.

Circuit board size and shape, then, is the second principal consideration. Of course the circuit complexity may well govern the size, but it is quite possible that a standard board outline such as a Eurocard (100 × 160 mm) will be specified (see Figure 9.3). If you have the luxury of 'breadboarding' your circuit first you will get a fair idea of its eventual size and shape. Figure 9.7 illustrates a typical odd shape which might be needed to fit around large components in some equipment.

Finally, you have to decide whether to use a single sided or double sided board. For a given complexity, single sided boards are harder to design but are less troublesome to make. They do have the disadvantage that jumper wires are often needed to replace difficult to place tracks, so are usually only used for relatively simple designs. It helps if these can be on the same pitch as your resistors, as component preforming tools can then be used to make them. Usually, once you have more than 25 or so jumper connections on a board it is time to think double sided. One trick frequently used in multimeters (where switches are arranged in banks) is to use a second small single sided board on top of the switches just for interconnections.

Basic layout considerations

It is a good idea to collect the actual components to be used (together with any alternatives if many products are to be made) so as to make sure that you appreciate the different physical sizes of each. Passive components such as resistors and capacitors come in a wide variety of shapes and sizes. They may have axial or radial leads. Axial leads project from either end of a component, whereas radial leads just extend from one

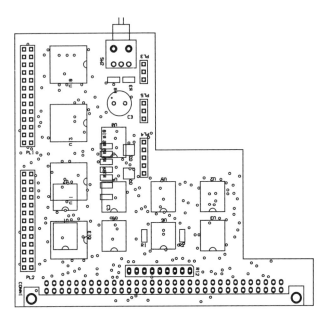

Figure 9.7

end. Axial leads have to be shaped or bent at right angles so that they can be passed through the PCB. Here the lead spacing or pitch is left to the designer. Radial leads, on the other hand, are at a fixed pitch and generally pass through the PCB without any shaping being necessary. Large axial components are, however, easier to support than radial ones (Figure 9.8).

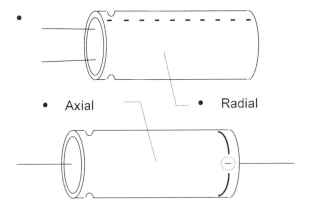

Figure 9.8

Other on-board components may be more complex and have very specific needs when it comes to assembly. Integrated circuits, in particular, have well defined outlines and fixed pitches, as do connectors and switches.

Every component lead must have its own pad or soldering point, which must be large enough to contain a hole for the lead and sufficient area surrounding the hole for a solder connection.

Knowing the body shape and size of each component is fundamental to the layout design process. It is very important that components are never so close to one another that they affect each other's performance. Knowing their lead spacing also helps to fix the positions of the pads. These are usually placed on multiples of a 0.1-inch grid, this size being chosen because almost all standard electronic components have a lead spacing of 0.1 inch.

Surface mount components have much smaller pitches, and only connect to one surface. The number of possible routes under their bodies is usually much smaller, and there is the added handicap that their pads cannot be used to change layer. You will usually need to use a freehand mode to connect to them. Don't put vias *in* the pads, and connect with a fairly narrow track (the surface tension of a fat one will 'rob' the component of necessary solder).

If you use EASY-PC Pro or a similar CAD package, you may well find that most of these component parameters have already been defined for you and are accessible through specific libraries, especially those for non-standard components. Alternatively, it should be possible for you to specify your own component and assign a special parameter set to it.

Component placement

The initial approach here is to place the principal components at designated places. For instance, connectors, controls and some of the other bulkier components may need siting at strategic places on the board. When designing a simple control amplifier for an audio system, you would want to put the volume control on the front edge of a PCB and the input/output connections at the rear. A PCB carrying a digital circuit might need an edge connector to interface it to a motherboard or back plane. IDC header plugs are frequently used as well. It is important that all these features are designed into the PCB at the start, rather than trying to add them later in the design process.

The regular components, such as resistors, capacitors and integrated circuits, should be placed in a logical order, usually broadly mapping the arrangement devised on the schematic diagram. This at least makes for easy troubleshoooting, but invariably wastes a lot of board space. Much 'tighter' layouts can be achieved by placing components parallel to one of the board's edges in fixed columns or rows. A digital circuit comprising more than 75% integrated circuits may well be easier to construct if those devices are placed first in several rows, with the other discrete devices added later (Figure 9.9). Memory boards and complex portable equipment (such as video cameras) provide the best examples of this.

Polarized components, such as electrolytic capacitors and diodes should be placed so that their directions of polarity are all the same. This makes component recognition very much easier and saves a great deal of trouble when it comes to assembly and final test time. Colour coded components such as resistors also fall into this category, as do those components which have their values marked on their bodies. Don't forget integrated circuits, which invariably have their pin-1 marked, sometimes by a notch in the top edge of the DIL moulding.

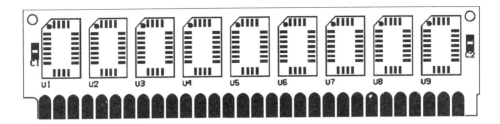

Figure 9.9

Whatever you do, it's a good idea to sketch your proposed layout first before committing yourself to anything more permanent. Draw the components in their desired positions and use this template as a guide. Unless you are about to draw your final layout by hand, where you have to place the pads first, a CAD package like EASY-PC Pro will allow you to capture your schematic drawing and present you with a 'rats-nest' of components, pads and tracks. Then it is just a case of moving each component (with its pads and connected tracks) to the desired place on the board.

Track routing

With all the components and their associated pads placed on the board, the next stage is to lay the tracks that connect the components to one another. You must take care not to allow adjacent tracks to come in contact with each other, or you will get a short circuit. It follows that one track cannot cross another without also making a connection. So if one route is barred by another track you will obviously have to reroute it around the offending track by laying the trace between the pads of a neighbouring component. Failing this, you may have to resort to a jumper wire that lifts the conductor off the board surface, over the other conductor (Figures 9.10 and 9.11).

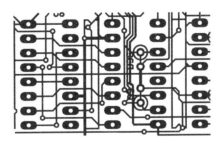

Figure 9.10 *Dense area showing track running between pads*

Tracks should normally be laid in straight or parallel lines and be of sufficient thickness to carry the required current. A thickness of 0.05 inch can handle about 4 A and is probably the thinnest usable width for power. With care, signal tracks can be as thin as 0.015 inch, but anything thinner may well dissolve during a prototype etching process. (Most PCB manufacturers can reliably produce tracks around half this thickness with their improved process control, while special techniques can produce

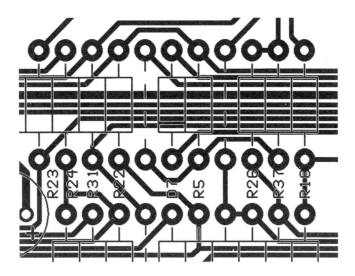

Figure 9.11

even finer tracks.) A track should not be wider than the diameter of the pad to which it is being connected. Making the track the same width can cause solder to flow away from the solder connection, but does provide better pad adhesion if the associated component is heavy or likely to be hot.

Parallel tracks should have a minimum clearance of 0.015 inch between each one (around 0.008 inch for manufacture); anything less may result in a short circuit. Bends in the tracks should be either 45° or 90° for neatness. Don't use acute angles, as capilliary action traps etchant, causing a 'well' to be etched away on the inside of the angle (Figure 9.12).

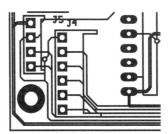

Figure 9.12

Routing tracks is largely a matter of common sense. Where double-sided boards are used, the normal convention is to run tracks in one direction on one side and at right-angles to this on the other side of the board. This keeps the hole count to a minimum and produces a reasonably well distributed conductor pattern.

Try to use the shortest possible route for conductors and aim to route from solder pad to solder pad. Stick to 90° (or greater) bends and avoid acute internal angles.

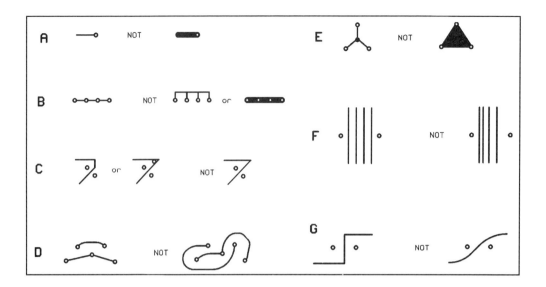

Figure 9.13

Figure 9.13 shows some practical 'dos and don'ts'.

A Avoid using conductors the same size as terminals as this will cause solder to flow away from terminal.

B Large multiple hole terminals may cause thermal sink when soldering.

C Sharp angles can cause capillary action that etches away track sections.

D Choose the shortest routing to avoid cross-talk.

E Stick to a regular pattern around holes to prevent solder fillets and difficulty in heating solder.

F Parallel tracks should be equally spaced to avoid build up of deposits when etching.

G This arrangement could also lead to shorted tracks.

Artwork identifiers

Identifying marks on the circuit layout are as important to the assemblers and testers as orienting the components in a standard way. These identifiers are added to the copper pattern; to indicate pin-1 of an integrated circuit or connector for example, or perhaps a polarity symbol for diodes and electrolytic capacitors, (Figure 9.14). Putting the circuit name and version number on the board somewhere is also a good idea. Note that, if this information is part of the copper on the underside, it must appear mirrored on your design. Remember you are working on a track pattern as viewed through the board from the top surface or component side.

There is a limit to the amount of this information, however. Too much extra copper may result in unwanted short circuits. One useful addition to a production-standard board would be a silk-screen pattern applied (usually) to the component side. Not only would this display the circuit name and board identification, but also the outline of all components, showing the circuit reference numbers and polarities. If space allows, the

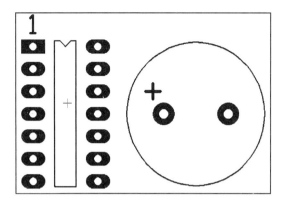

Figure 9.14

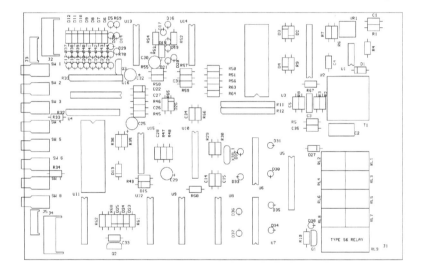

Figure 9.15

circuit reference numbers are better placed outside where the component rests, so they can be read on the finished board (for troubleshooting or diagnostics) (Figure 9.15).

Practical design considerations

Once the basic considerations have been appreciated, the proficient engineer must attend to certain design refinements. Remember that a PCB is a component in its own right, and layout is particularly important for very high frequency and fast digital circuits. The PCB designer must be aware of the transmission line effects caused by the copper tracks, so choice of track widths is important, as is the elimination of undesirable features such as ringing, cross-talk and supply line noise.

Track width

We saw earlier that a practical minimum width of power track would carry about 4 A. Even a width of 0.015 inch will carry a sustained current of over 1 A. Most of the conductors in a digital circuit carry small signal currents for which the conductor resistance can virtually be ignored, so narrow tracks are ideal. The supply and ground lines, however ought to be much wider than the signal lines, in order to improve their high frequency performance and reduces the voltage spikes that can appear on them due to the high current.

The ground potential must be stable, so the width of the ground conductor should be as large as possible, especially for TTL. For any circuit, a useful rule of thumb could be to make the ground width greater than the supply width, which itself should be greater than the signal width. As higher application speeds and currents are required, you will need to scale these widths up significantly. It is also important to cross-connect power routes as much as possible, so that a surge can be drawn by as many paths as possible. This reduces its depth, and ensures that any differential in signal levels is minimized (as more devices suffer a smaller pulse).

Careful positioning of ground and supply tracks can provide an element of transmission line capacitance that can be used for decoupling. If a double sided board is being used these tracks should be placed directly opposite one another on either surface for the same effect, but only, of course, where this does not interfere unduly with available routes. On a practical note, if instead they are run parallel to each other, between the rows of IC legs and immediately under the ICs, almost all the signal tracks will then be visible on an assembled board for fault tracing. On a single sided board, decoupling can be achieved by positioning tracks adjacent to one another in a similar manner.

For TTL circuits, designers should choose signal track widths such that the transmission lines created have a characteristic impedance of $100-150\,\Omega$. This means that the track widths should be as small as possible, typically half as thick as the board itself. Wide signal tracks result in low impedance lines, which are very susceptible to current spikes.

For CMOS circuits, signal line widths are less critical. CMOS circuits have much higher input impedances and their tracks have higher characteristic impedances.

The minimum recommended spacing for tracks is the same as the conductor width, but a more generous spacing should be used where practical. This will reduce interline capacitance and improves the reliability of the manufacturing process.

Ringing

Ringing is created in a circuit when the input and output impedances of the gates do not match the characteristic impedance of the transmission lines created by their interconnections. A simple solution of the problem is difficult because the impedances of these gates are non-linear. When lines between components are kept very short the problem is virtually eliminated. Where long lines are unavoidable the effect can be minimized by adding compensating resistors. Adding buffers or line drivers to the circuit is another possibility, but runs the risk of increasing radiation from the supply lines due to current surges.

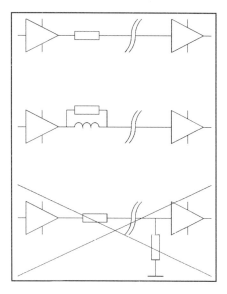

Figure 9.16

A resistor connected in series with the output of a driving gate (Figure 9.16) may improve the trailing edge response of the signal. Its value added to the output impedance (a.c.) of the driving device should roughly equal the characteristic impedance of the line. With TTL the current induced voltage drop may be too great. In this case a series inductor can be used. Either a lossy core, or a shunt resistor is needed for damping. The values depend on the line capacitance in each individual case, but any resonance should be at least critically damped, and preferably overdamped, if the increased propagation delay is to be tolerated. Resistors at the receiving end can also improve rising and falling edge performance but, because they present a current drain, they should not be used except with specialized line drivers (using non-TTL signal levels) designed to provide the increased input signal required.

For long lines, using line drivers is probably the best solution, in spite of the additional cost. Drivers are usually available as single IC transceiver packages that can be used at either the sending or receiving ends of a line. They have controlled rise and fall times to reduce harmonics, and match the impedance of the lines to eliminate reflections.

Cross-talk

Cross-talk is more prevalent in TTL circuits than CMOS circuits, which have better noise immunity. Designers now frequently choose CMOS, as much higher speeds are achieved with the newer families of this technology. There are a number of approaches:

- If TTL is used, keep adjacent signal tracks as far apart as possible, but close enough to a ground plane to make use of the decoupling effect of the line capacitance.
- Where there are parallel lines, such as buses, make sure that the signals travel in the same direction. If this is impractical, add a ground track between the signal lines.

Supply noise

Noise carried along both the supply and the ground lines represents another serious problem for designers, especially in digital circuits. Current spikes occur at the output of a logic gate every time there is a transition through it. Although this current spike has a very short duration, typically 5 ns for a TTL gate, it can be as high as 25 mA. An additional spike occurs where the logic state at the output of a gate changes. This is caused by the charge or discharge of the transmission line capacitance, and could contribute a further 20 mA in an extreme case.

The cumulative effect of these individual currents represents a significant drain on the power source. A digital IC having 20 logic gates could require an extra 500 mA, which appears as a transient voltage drop (Figure 9.17) across the device because of line impedance. To overcome this problem the track layout designer could provide a ground plane with as much copper surface as possible. A large surface like this would have very low resistance and provide some electromagnetic screening. Provide for wide supply line tracks on the board – typically 5–10 mm. Site the supply track as close as possible to the ground plane, ideally on the opposite side of a double sided board. Doing this provides a low characteristic impedance between V_{CC} and ground.

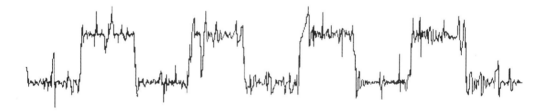

Figure 9.17

It is a good idea, and often imperative, to arrange separate ground lines for the digital and analogue areas of a circuit. This also applies to input signal and output signal ground lines in analogue equipment. These separate ground connections should be taken to a common earth point in the equipment. This is vital with analogue-to-digital converters, and vice versa.

Finally, make sure that you allow for decoupling capacitors throughout the circuit – at least one for every two or three ICs. These should be either ceramic disk or polyester multi-layer plate types of about 10 nF for TTL and 5 nF for CMOS. Do not use wound polystyrene capacitors because these introduce an undesirable inductive component. The variety using sandwiched interdigitated plates has been shown in studies to perform as well as ceramic types up to high radiofrequencies. In extremely fast circuits, containing significant power in the gigahertz region, lead inductance becomes significant. Surface mounted capacitors are then the most practical option, but exact positioning becomes critical.

Practical Exercise: printed circuit board design

This exercise will take you through all the stages necessary to produce a small electronic device, starting with the schematic circuit, then simulating it, translating the circuit to a PCB layout, and ending with circuit assembly and test.

The circuit you are to make is a logic pulser. This very handy device is used to inject signals at chosen points in a digital circuit, and normally comes with a companion tool called a logic probe, which detects the pulses elsewhere in the circuit. A schematic diagram of the circuit is shown in Figure 9.18.

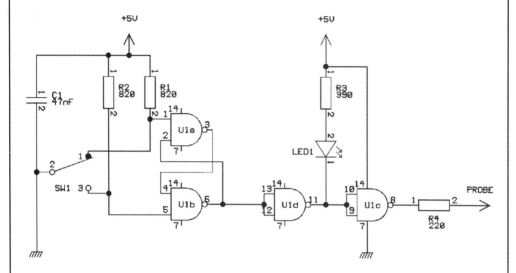

Figure 9.18

Operation

The circuit is based upon the simple RS latch arrangement which we studied in Chapter 5. This time we use a 74LS00, which is a TTL integrated circuit that contains four 2NAND gates. We only need to use two of these gates but, because we should not leave the inputs to the other two floating, we connect them as simple inverters. U1a and U1b are cross-coupled and deliver a positive pulse at the output of U1b when the switch SW1 is pressed. This switch is single pole, double throw (SPDT) and spring loaded. Pressing the switch delivers a RESET signal and when released the spring returns the contacts to the initial SET position.

On RESET the low output at pin 11, as well as being presented as a high at the logic pulser tip, turns on the LED. On SET the output is high and the LED will be off. The LED is connected to +5 V as LS TTL will sink 8 mA, but only source

Continued on p. 214

Practical Exercise: printed circuit board design *(Continued)*

400 µA. Note that if HCMOS were used, it can only sink or source 4 mA, so R3 would have to be increased to 820 Ω, and a low current LED used.

The resistors R1 and R2 are pull-up resistors to prevent either disconnected gate input from floating. R3 defines the current through the LED and R4 defines the circuit output impedance. C1 decouples the circuit, as in practice it is supplied by flying leads. IC1c and IC1d are swapped so that the connected pins are physically adjacent.

Although the main idea of this exercise is to take you through the process of producing a printed circuit assembly, the final mechanical production will be left to you. Conceptually the product needs to be in some sort of cylindrical housing, large enough to be hand-held, with the switch easily accessible and the LED clearly visible. The probe end needs to project roughly 1 inch out of the end of the assembly. The typical logic pulser shown in Figure 9.19 should give you some idea.

Figure 9.19

Schematic

Step 1: Start EASY-PC Pro and select schematic: [A].

Step 2: Check the status line and switch to Zoom4: [4].

Step 3: Starting from roughly the centre of the drawing area, place a 7400 (Quad 2NAND): [F8], [B], 74LS.IDX, 74LS00.

Step 4: Check that the details in the SCH Placement Reference are correct: [K] to continue.

Step 5: You should have four 2NAND sections on the screen: press [ESC] to complete the placement.

Step 6: Put the cursor between U1b and U1c, and press P to pan to centre: all four sections should now be visible.

Step 7: Switch off the pin names by keying SHIFT F2, SHIFT N; the SHIFT F2 combination brings up the Preferences menu (Figure 9.20). Then ESC back to drawing screen.

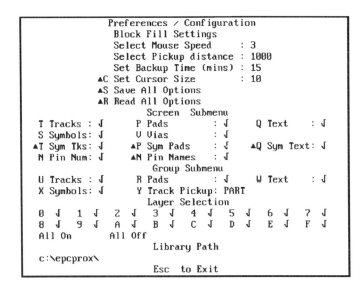

```
              Preferences / Configuration
                Block Fill Settings
                Select Mouse Speed      : 3
                Select Pickup distance  : 1000
                Set Backup Time (mins)  : 15
         ▲C Set Cursor Size             : 10
         ▲S Save All Options
         ▲R Read All Options
                    Screen   Submenu
   T Tracks : √       P Pads        : √     Q Text     : √
   S Symbols: √       U Vias        : √
  ▲T Sym Tks: √      ▲P Sym Pads    : √    ▲Q Sym Text: √
   N Pin Num: √      ▲N Pin Names   : √
                    Group Submenu
   U Tracks : √       R Pads        : √     W Text     : √
   X Symbols: √       Y Track Pickup: PART
                    Layer Selection
   0 √   1 √   2 √   3 √   4 √   5 √   6 √   7 √
   8 √   9 √   A √   B √   C √   D √   E √   F √
   All On        All Off
                    Library Path
   c:\epcprox\
                   Esc  to Exit
```

Figure 9.20

Step 8: Arrange the 2NANDs as in Figure 9.18, making sure that each section is positioned as shown. Place the cursor near pin 9; F7 makes it mobile; move it to the right about 3 inches, then ESC.

Step 9: Repeat for pin 12. There is no need to key F7 again, just click, as component EDIT is still active. Place the cursor over pin 12 and click the left-hand mouse button. Move this gate element 1.5 inches to the right.

Step 10: Repeat for pins 4 and 1, placing gate a directly above gate b.

Step 11: Using New Line (F2), connect up all the pins. You will need to switch to lines at 90°: SHIFT A, B. Note the 'beeps' as connections are made.

Step 12: Draw short line stubs on pins 7 and 14 of U1c, giving these the net names GND and VCC, respectively: SHIFT N, etc.

Step 13: Now place resistors R1 to R4; deal with R1 first. Put the cursor near (not on) pin 1, then key F8, B, PULSAR.IDX and select RESISTOR.

Step 14: On the SCH placement click on Value 1 | R:1K.

Step 15: On the Values screen, change R:1K to R:820. Note that the R: part is very important. Use K to accept and K to place (Figure 9.21).

Continued on p. 216

Practical Exercise: printed circuit board design *(Continued)*

```
SCH Placement Reference
Reference  I  R              I  √
Number     I  1
Component  I  RESISTOR       I  .
Package    I  DSC            I  .
Value 1    I  R:820          I  √
Value 2    I                 I  .
Values     I       more
              OK
```

Figure 9.21

Step 16: Shift the component until it is one grid away from pin 1 – but do not press `ESC`.

Step 17: Move the cursor left about 0.4 inch and key `R` to repeat.

Step 18: Repeat again above pin 11 (we'll change the value in a minute) and repeat near pin 8, but this time rotate through 90°: `A` `1`.

Step 19: `SHIFT` `P` to bring up the placement reference menu: for this last resistor, change the value to R:330

Step 20: `K` to accept and place, followed by `ESC` to release.

Step 21: Take the cursor to the resistor near pin 11 and key `F7` to edit; change this value to R:390.

Step 22: Now add the LED to your circuit; use the LED which is provided for you in the DISCRETE.IDX library. Similarly, place a 47 nF capacitor adjacent to R2, using C from DISCRETE.IDX. Use New Line (`F2`) to connect the LED to R3 and connect the cathode to pin 11 on IC1d.

Step 23: Connect the change-over switch; use SPDT in the DISCRETE.IDX library. We need to use a vertical component, so before it is fixed in place make sure that the Package type is set to DSCV (Figure 9.22). Use New Line (`F2`) to connect the switch to R1 and R2 and then onto the two gates; connect its common end to the GND net and the bottom of C1.

```
SCH Placement Reference
Reference  I  SW             I  √
Number     I  1
Component  I  SPDT           I  .
Package    I  DSCV           I  .
Value 1    I                 I  .
Value 2    I                 I  .
Values     I       more
              OK
```

Figure 9.22

Step 24: Link the other ends of R1, R2 and C1; call the net VCC, and extend the VCC stub on IC1c pin 14 to connect to the top of R3.

Step 25: Add a short stub to the unconnected end of R4 and call this net OUT.

Step 26: To finish off the circuit, add schematic symbols for the supply, ground and output (use +5 V, EARTH and TO from the SCHEMA.IDX library). Make sure that each symbol is connected to its appropriate net – check this by picking up each symbol (F7) and placing it elsewhere on the diagram. If the track is taken as well then you have a connection.

Step 27: Save the circuit as PULSER.SCH: F9 S , etc.

Simulate in PULSAR

Step 28: Because the switch is not an active component we must emulate it somehow. Name the nets on pins 1 and 5 as PIN1 and PIN5. Now select Logic – PULSAR from the Tools menu to bring up PULSAR.

Step 29: Apply a signal (20 kHz) to PIN5 and then apply the same signal, but 180° out of phase to PIN1.

Step 30: You can create such a signal using Generator. Load 20 kHz.GEN, making sure the cursor is at the beginning of the trace; insert a strong high state of duration 25 μs (the same as that indicated in the Time panel). Save this generator as 20K_1.GEN (or anything suitable to identify it, but don't use the characters 'Hz' in the filename).

Step 31: Return to PULSAR and simulate; you should get a trace like the one in Figure 9.23, which shows the output changing as the inputs change. 'OUT' is blue to indicate that it is an overrideable signal, as it is fed through a resistor.

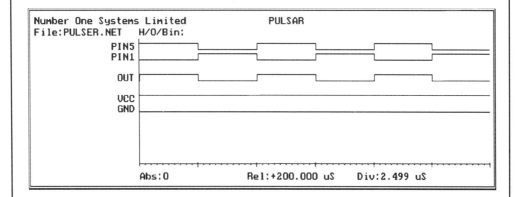

Figure 9.23

Step 32: Quit PULSAR and return to the EASY-PC Pro schematic.

Continued on p. 218

Practical Exercise: printed circuit board design (*Continued*)

Output

Now that your schematic is complete and the circuit has been simulated satisfactorily, you are ready to proceed with the design of the PCB layout. Before doing this it is good idea to make a 'hard-copy' of your drawing, just for the record. Refer to Chapter 6 for guidance on how to do this. Note Figure 6.31 for the output options, and follow steps 1–19 in Chapter 6.

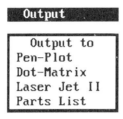

Figure 6.31

Parts list

This is also an ideal point to create a parts list. Choose the Parts List option offered in the output menu and save your list to File (Figure 9.24). The parts list will be saved to an ASCII file called PULSER.LST which can be used later to add the finer details concerning your finished circuit. Don't forget to save your current schematic.

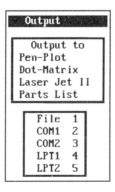

Figure 9.24

Translation to PCB

Step 33:   From the Tools menu choose Translate to PCB (Figure 9.25). After a moment the program will have switched from SCHEMATIC to PCB EDIT

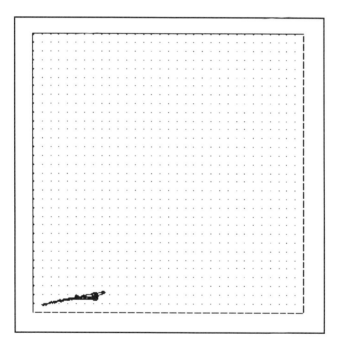

```
┌─────────────────────────────────────────┐
│ View        Tools    Settings           │
└─────────────────────────────────────────┘
        ┌───────────────────────────────┐
        │          Simulators           │
        │ Analogue  -  Analyser III      │
        │ Logic     -  Pulsar            │
        │ ─────────────────────────────  │
        │ Junction Dots                  │
        │ ─────────────────────────────  │
        │ Translate to PCB               │
        └───────────────────────────────┘
```

Figure 9.25

Figure 9.26

mode, and you end up with a screen like the one shown in Figure 9.26. (You will get a message 'Package not found', as the labels you have used have no PCB equivalent – pressing ESC clears the message.)

Step 34: All your schematic components connected in the net-list (except for the labels) now appear at the bottom left-hand side of the drawing area. Put the cursor in the middle of the 'string' and go to zoom 5 (Figure 9.27). The connections should be quite clear now, and all you have to do is to

Continued on p. 220

Practical Exercise: printed circuit board design (*Continued*)

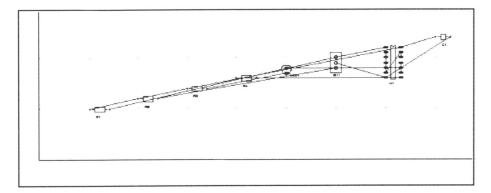

Figure 9.27

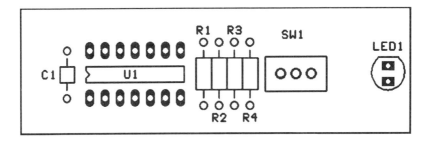

Figure 9.28

arrange the components into the required configuration. Redraw screen ⌨ALT⌨ ⌨R⌨ can help here. Use the outline shown in Figure 9.28 as a guide.

Step 35: Use ⌨F7⌨ to pick up each component in turn and place it somewhere away from the corner of the drawing area. Note that the tracks extend, as they keep their connections to other components. The red track is a stub and is connected in this case to the output pin. Figure 9.29 gives some idea of the intermediate stages.

Step 36: With your components placed in their new positions, open the Nets menu, shown in Figure 9.30, and optimize all nets: ⌨ALT⌨ ⌨G⌨. This powerful feature checks the current circuit arrangement and re-draws all connections, making sure that the shortest route is used in each case.

Step 37: 'Fine-tune' your placements and re-optimize until your layout looks roughly like the one in Figure 9.31.

Step 38: Use ⌨F5⌨ to put all component reference numbers within each component outline.

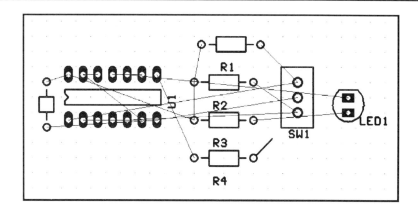

Figure 9.29

```
┌──────────────────────────────────────┐
│ Tools      Nets      Settings        │
├──────────────────────────────────────┤
│ Show Net             Alt S           │
│ Hide Net             Alt H           │
│ Show Next Net        Alt N           │
│ Show Previous Net    Alt P           │
│        Optimise Nets                 │
│ Current Net          Alt O           │
│ Component            Alt C           │
│ All Nets             Alt G           │
└──────────────────────────────────────┘
```

Figure 9.30

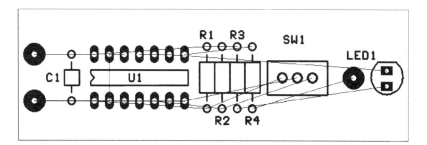

Figure 9.31

Now we have to turn these net-list connections into tracks, trying to keep as many as possible on the opposite side of the board to the components. Before moving on you may need to review what you have learnt about 'line editing' with EASY-PC Pro.

Continued on p. 222

Practical Exercise: printed circuit board design *(Continued)*

Track or line edit

You can access track edit by selecting Edit Track from the Edit menu (Figure 9.32). Alternatively, keying [F1] will 'pick up' the track node nearest to the cursor. All the available track editing features can be accessed by clicking on the Track menu which appears. These options are virtually the same as those used on schematic line edit, with some additions (Figure 9.33).

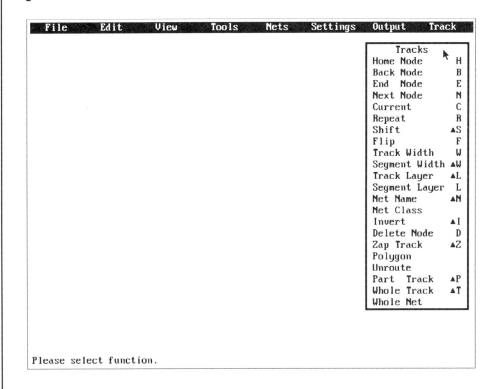

File	Edit	View
Edit Track		F1
New Track		F2
Edit Pad		F3
New Pad		F4
Edit Text		F5
New Text		F6
Edit Component		F7
New Component		F8
Group Ops		F10

Figure 9.32

File   Edit   View   Tools   Nets   Settings   Output   Track

```
              Tracks
    Home Node      H
    Back Node      B
    End  Node      E
    Next Node      N
    Current        C
    Repeat         R
    Shift         ▲S
    Flip           F
    Track Width    W
    Segment Width ▲W
    Track Layer   ▲L
    Segment Layer  L
    Net Name      ▲N
    Net Class
    Invert        ▲I
    Delete Node    D
    Zap Track     ▲Z
    Polygon
    Unroute
    Part  Track   ▲P
    Whole Track   ▲T
    Whole Net
```

Please select function.

Figure 9.33

As before, each track has a home and an end position and comprises two or more nodes or connection points along its length. When you move the cursor during Line Edit the track becomes mobile and is dragged along with the cursor – this is known as *rubber banding*. Now you can move the track between obstacles and lay it in any desired position.

Each time you click the left-hand mouse button you anchor another node to the track, from which you can branch away in a new direction. You may have to switch between 45° and 90° angled tracks for some routes, and when you position tracks between the IC pads remember to make that track segment much smaller: ⎡SHIFT⎤ ⎡W⎤ ⎡2⎤. If you pick up the wrong track by mistake, select Unroute, Whole Track from the Line menu to cancel the selection, or ⎡SHIFT⎤ ⎡T⎤.

Laying down PCB tracks

Step 39: Snap to quarter grid: ⎡M⎤ ⎡Q⎤. Start with the track from R4 to the IC pin 8. Place the cursor at the resistor end and key Line Edit: ⎡F1⎤. The track changes to red, which indicates that it is on the upper side or layer 1.

You can verify this by inspecting the status bar – keying ⎡+⎤ will reveal the right-hand end. Change the layer to the underside of the board by keying ⎡SHIFT⎤ ⎡L⎤ ⎡E⎤. The track will change from red to blue. Make the track size 0.25 inch: ⎡W⎤ ⎡5⎤.

Step 40: When you are satisfied with its position, press ⎡ESC⎤ to end the edit.

Step 41: Now continue to 'lay down' the rest of the tracks. You will see that status bar has a red 'l', indicating that Line Edit is still active, so there is no need to key ⎡F1⎤ again. Just click the left-hand mouse button on the desired track to make it active. If you pick up the wrong track, refer to the next section to find how to recover.

Step 42: Use ⎡F4⎤ to add three extra pad roundels (shown in Figure 9.31 and Figure 9.34 for the supply and output connections). Make their width

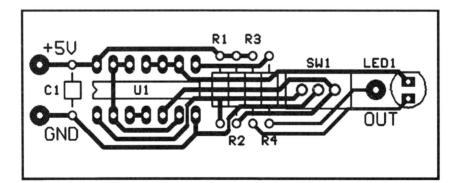

Figure 9.34

Continued on p. 224

Practical Exercise: printed circuit board design (*Continued*)

size 8 (W   8) and the hole size 2 (H   2). Anchor the supply, ground and output tracks to these three points. We may have to change the shape and position of the output pad later.

Step 43:   Leave the GND track until last as it may prove to be the most troublesome to lay down.

The finished layout could look something like the one in Figure 9.34.

Once the layout is completed there are certain checks that can be done to ensure a trouble-free product.

Integrity checking

This compares the original net-list with the final layout and displays a message on the screen when there is a conflict. As the +5 V, EARTH and TO components have no matching PCB part, you will get errors.

Two new report files are created: PULSER.PDN and PULSER.SDN. The PDN net-list shows all the components and connections that exist on the PCB layout but not on the schematic, and the SDN file shows those that appear on the schematic drawing but not on the PCB. You will need to load both these files into a viewer to establish whether or not the differences are acceptable.

Step 44:   Bring up the Tools menu shown in Figure 9.30 and select Integrity Check.

In this example we have added three roundels or pads for connections to the 'outside world' during the layout stage. The Integrity Check will not spot this as pads do not appear in the net-lists, but it will spot any deviations from the original schematic net-list.

Design rule check

Design Rule Check ensures that there are no clearance infringements in your layout.

Step 45:   Bring up the Tools menu again and choose Design Rule Check or just key ALT D . This gives you the design options menu (Figure 9.35).
Step 46:   Before you start the check you may wish to inspect the clearance rules themselves. Do this by keying c : this gives the default settings (Figure 9.36) – and you may wish to adjust some of them later on.
Step 47:   Press ESC to return to Figure 9.35 and start the design rules check by keying S . 'EasyPC Pro' checks through your layout and, when

```
Design Rule Options
C Setup Clearances
S Start Checking
H Clear Highlight
1 Track to Track ON
2 Track to Via   ON
3 Via to Via     ON
4 Track to Pad   ON
5 Via to Pad     ON
6 Pad to Pad     ON
```

Figure 9.35

```
Design Rule Clearances
1 Track To Track   10
2 Track To Pad     10
3 Pad    To Pad    15
```

Figure 9.36

```
          Design Rule
   Error(s)
     Track <-> Track    8
     Track <-> Via      0
       Via <-> Via      0
     Track <-> Pad      2
       Via <-> Pad      0
       Pad <-> Pad      0
Press any key
```

Figure 9.37

completed, presents you with a report of any failures (Figure 9.37). Ten errors are flagged in the Figure: eight track to track and two track to pad. As well as the Design Rule report the suspect areas on your layout are highlighted with an error message graphic (Figure 9.38).

Step 48: Choosing DRC Summary in the Tools menu, or [ALT] [U], will display the error messages again. This version provides a legend that describes each error message graphic (Figure 9.39).

Step 49: [ALT] [T] or Next DRC Error moves to the next problem (Figure 9.40).

Although they are not showing an error, it would be preferable to include the extra three connection pads as components in the schematic drawing. This would entail designing the symbols, building the components and adding them to your libraries.

Continued on p. 226

Practical Exercise: printed circuit board design *(Continued)*

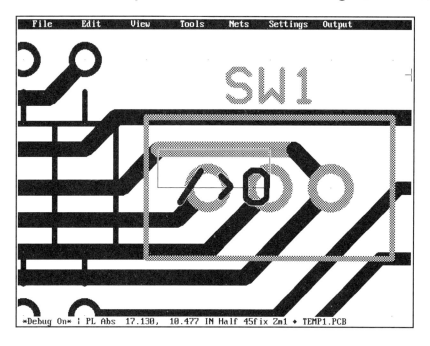

| File | Edit | View | Tools | Nets | Settings | Output |

SW1

Debug On | PL Abs 17.130, 10.477 IN Half 45fix 2m1 ♦ TEMP1.PCB

Figure 9.38

```
DRC Summary
Unrouted Tracks  /-?  1
Track to Track   /-/  7
Track to Via     /-o  0
Via to Via       o-o  0
Track to Pad     /-0  2
Via to Pad       o-0  0
Pad to Pad       0-0  0
Total                 10
─────────────────────────
√   Display Errors    D
─────────────────────────
             OK
```

Figure 9.39

Step 50: Make any other corrections that are necessary, and re-run the check. You may wish to turn selected options on or off (Figure 9.35), or even adjust clearance settings, (Figure 9.36), but alter these with caution.

To complete the layout we must add some extra details. First let us consider the output probe connection. This will entail a slight readjustment to your layout.

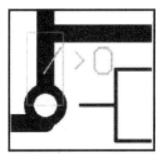

Figure 9.40

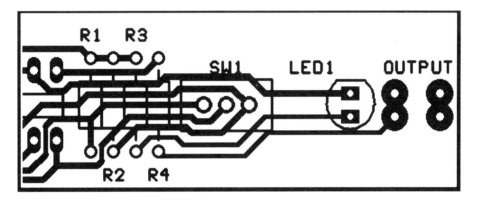

Figure 9.41

Specifically, the tracks from the lower end of R3 and R4 need to be interchanged to allow the output to reach the tip. You will find that clearances are tight, and several extra segments will be needed in the tracks between the resistors to maintain them (see Figure 9.41).

Step 51: Locate the 'output' roundel and move it to the right of the LED. `F3` will allow you to pick up the pad and, as you relocate it, the track will remain attached and 'rubber band' to its new position.

Step 52: Do not press `ESC` just yet, but repeat (`R`) the roundel three more times until you get an arrangement like the one shown in Figure 9.41.

Step 53: Use New Line (`F2`) to draw an outline around the four roundels. The rectangle must be a complete net, but make sure that the track does not join the out net.Enlarging the cursor, `SHIFT` `C`, might help here (Figure 9.42).

Step 54: Go to the track edit menu (Figure 9.33) and select Fill Polygon. The rectangle space should now flood-fill (Figure 9.43). Don't worry if the roundel holes disappear – they are restored the next time the screen is

Continued on p. 228

Practical Exercise: printed circuit board design *(Continued)*

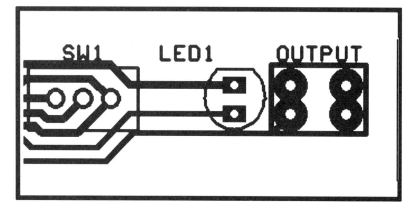

Figure 9.42

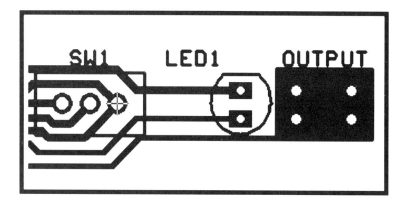

Figure 9.43

refreshed. Whatever you choose as a probe, contact can now be fixed to this region.

Finally, we need to mark the boards cropping points and write a brief description or reference number on the copper side.

Step 55:   Add two right-angled corners to the left-hand side of the board. The inside edges are the cropping guides and should align just outside the outside edges of the two tracks that run the length of the board. (It is good practice to leave a narrow insulating border around the edge of a board – Figure 9.44.)

Step 56:   Repeat the crop marks on the right-hand side so that the board can be cut to a rectangle. Then add the additional marks so that the probe end

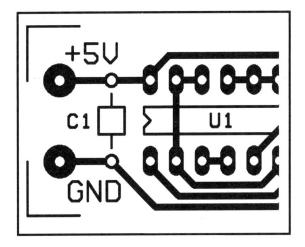

Figure 9.44

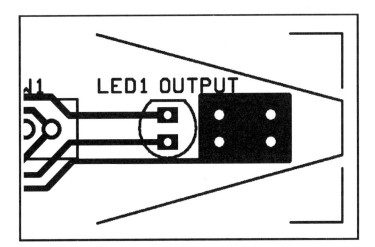

Figure 9.45

can be shaped (Figure 9.45). You will have to use multi-angled track (⌨SHIFT A M) to draw this shape. To complete the copper side you will need to add a circuit name or reference. The best place for this text is at a corner or edge of the board. There is a large space just above the switch Figure 9.46.

Step 57: Position the cursor above the switch and key ⌨F6 to enter the Text mode. Type the word PULSER in the panel that appears in the top left-hand corner of the screen and press ⌨ENTER to transfer text to the drawing.

Continued on p. 230

Practical Exercise: printed circuit board design *(Continued)*

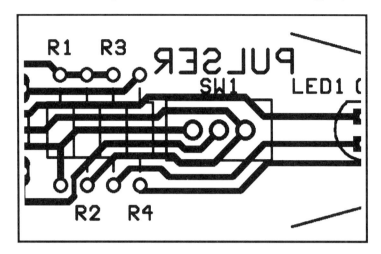

Figure 9.46

Step 58:   Make sure that the text appears on layer E ([L] [E]) and is of size 2 ([S] [2]) and width 2 ([W] [2]).

Step 59:   Remember that what you are seeing on the screen is a component or top-side view of the tracks. The copper side is going to be a mirror image of this. So, before you fix the PULSER text into position, you must flip it ([F]) so that it appears mirrored looking from the top side of the board.

Step 60:   [ESC] will complete the Text placement.

PCB production

Outputting the layout

The method for taking a hard-copy of your printed circuit layout is virtually the same as that for schematic diagrams. This was described in Chapter 6, so it might be a good idea to familiarize yourself with the procedure described in steps 1–19 in that chapter.

First, we generate a draft copy of the layout.

Step 61:   Use [F10] to select the area to be printed (Figure 9.47). (*Note*: If no area is specified, EASY-PC Pro will automatically detect the size of the design, and print just this area.)

Step 62:   Access the output menu. This looks the same as the schematic output menu, but with a few additions (Figure 9.48).

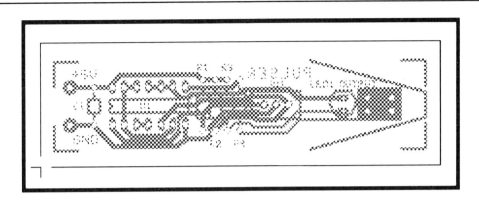

Figure 9.47

```
┌──────────────────────────────┐
│ Output                       │
├──────────────────────────────┤
│     Output to                │
│   Pen-Plot                   │
│   Dot-Matrix                 │
│   Laser Jet II               │
│   Gerber                     │
│   NC Drill                   │
│   Parts List                 │
│     Export                   │
│   DXF                        │
│   Stockit                    │
└──────────────────────────────┘
```

Figure 9.48

Step 63: Choose the output option most suited to your installation. Compare the Laser Jet menu shown in Figure 9.49 with that illustrated in Figure 6.32. Figure 9.49 highlights the slightly different settings for PCB production.

Step 64: Steps 6–19 in Chapter 6 describe most of the settings. Printed circuit layouts, however, require attention to a few exceptions.

Step 65: The layers must be output Separately not Together. Either key Ⓛ, or click the mouse on Layers Output to change it.

Step 66: For a draft-quality printout, key Ⓜ to reduce the resolution from 300 dpi (dots per inch); 75 dpi is quite adequate here.

Step 67: Ensure that the scale is 1:1 for full size output. The full size Print should fit horizontally onto an A4 sheet this time, although for larger layouts in the future you may have to rotate the highlighted image through 90° first. Use Group Rotate for this.

Continued on p. 232

Practical Exercise: printed circuit board design *(Continued)*

```
            EASY-PC Professional, Laser printer output
   Layer
 0 Silk      On                   I Input From    : TMPFILE.PCB
 1 Copper    On    Resist (.LUx)  O Output To      : LPT1
 2 Copper    Off                  L Layers output  : Separately
 3 Copper    Off                 ▲O Solder resist  : No
 4 Copper    Off                  H Pad Holes      : Avoid
 5 Copper    Off                  M Resolution     : 150 dpi
 6 Copper    Off                  P Paper          : A4
 7 Copper    Off                  N Copies         : 1
 8 Copper    Off                 ▲E Scale          : 1.000
 9 Copper    Off                 ▲F Print from     : 2.100,3.700   in
 A Copper    Off                  T Print to       : 5.450,4.600   in
 B Copper    Off                 ▲P Print offset   : 0.000,0.000   in
 C Copper    Off                  G Pin names      : Off
 D Copper    Off                 ▲G Pin numbers    : Off
 E Copper    On                  ▲A Compensation ▲I Summary
 F Silk      Off                 ▲C Centre Print  S Start Print
                                 ▲S Save Setup    R Restore Setup
   Print will FIT                 U Units         Q Quit
```

Figure 9.49

Step 68: Centre the print: [SHIFT] [C].

Step 69: Key [S] to start.

Step 70: When you are happy with the product, repeat the procedure from step 61, but this time increase the resolution to, say, 300 dpi.

Step 71: Printing the image onto clear acetate for transfer to copper-clad circuit board is often recommended. However, materials that hold a static charge often give poorer contrast with laser printers, as toner transfer is not so efficient. The result is pinholing, or porous etch resist. Using a good quality translucent draughting paper will often give the best results. Alternatively, for a low-tech solution, lightweight paper (60 gsm or lower) can be made transparent by using a special spray. Do not use draughting film in a laser printer, unless you like cleaning sticky messes from the fuser unit!

Preparation for production

We won't cover the process of manufacture here as there are many methods, each of which depend upon actual choice of facilities. In addition, it is not a practical proposition for the single user or for the one-off product. At the very least it would involve buying a light-box, chemical solutions, an etching bath, all manner of accessories, not to mention PCB material itself. These may be a worth-while investment if you intend to make a great many PCBs, bearing in mind that several of the chemicals have a limited shelf-life. Many schools and colleges will have their own PCB fabrication facilities, but for small scale or home use, one of the (usually mail order) disposable kits which carry out processing inside a sturdy polythene bag may be preferable. Modern etch resist coatings are usually fairly resistant to tungsten (filament) light, but beware fluorescent lights and daylight. If a light-box is not available, two sheets of thin glass, strong sunlight and a degree of experimentation will work.

Third party production

Assuming that a 'third party' will make the board, you ought to obtain some infor-mation before starting the design. You need to know from the manufacturer:

- the maximum board size handled;
- the maximum number of layers;
- the minimum track width;
- the minimum silk-screen line widths;
- the minimum clearances allowed between track/track, track/pad, pad/pad, etc.;
- the scale of artwork required, or the format of disk files if preferred; and
- the normal manufacturing tolerances.

  You should always provide the manufacturer with a mechanical drawing of your board as well as all the artwork. A mechanical drawing is an often neglected feature in PCB design. This drawing should contain all the physical attributes of the board: its outline size and shape, the position and sizes of mounting holes, and the board

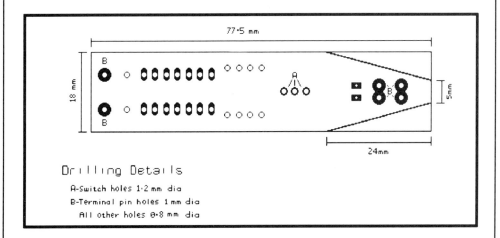

Figure 9.50

Continued on p. 234

Practical Exercise: printed circuit board design *(Continued)*

thickness and composition. In addition, you should specify the number and size of all fixing holes, vias, plated through holes, as well as any other relevant information that the PCB manufacturer should know. All these points can be draughted onto one of the other drawing layers. Tolerances will usually only need to be included where they are critical, or abnormally tight. This information provides the basis for any quotation if fabrication is to be placed with a commercial manufacturer, especially if several boards are to be made and, of course, it is essential if special tooling is required.

Figure 9.50 shows part of a mechanical drawing that you might produce for the PULSER circuit. It would be more commercially acceptable if the drawing were framed properly, and included an author's name and document management details. Figure 9.51 is a suggestion of what might be an acceptable drawing.

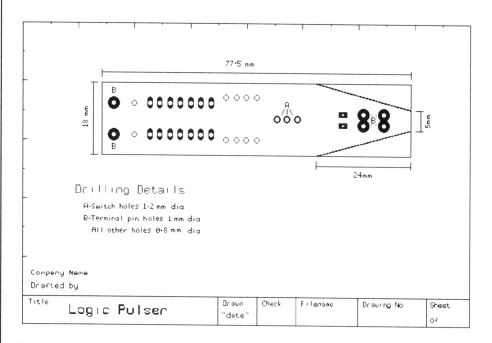

Figure 9.51

10 Digital electronic projects

In this final chapter the reader is offered a variety of simple circuits to investigate. All schematics can be drawn using EASY-PC Pro and most can be simulated using PULSAR. However, some of the components used are not included in the supplied libraries, and this is the chapters challenge. You will need to create schematic symbols and components as well as PCB symbols and components as necessary. Look out for seven-segment displays and phototransistors.

1. Logic probe

This device will make a useful companion for the logic pulser developed in the last chapter. The logic pulser is used for injecting a signal at some point in a digital circuit, and the logic probe will sample the data at some suitable point elsewhere in the circuit and indicate whether a logic 0 or a logic 1 exists. Both devices are normally supplied in simple cylindrical housings, not much larger than a cigar case. The probe end is normally some hard-wearing metal spike, something like a 1-inch nail, whereas the other end has two flying leads that are connected to a convenient 5-V source.

The logic probe described here (Figure 10.1) comprises a single 14-pin IC package that contains six inverters – a 74LS04, and we need to use five of them. The input of the unused section must be tied high, even though unused TTL inputs tend to drift high. Two LEDS indicate logic 0 and logic 1: the red LED indicates high and the green LED indicates low. Both LEDs require at least 15 mA to turn on, so we must connect two pairs of inverters in parallel to give sufficient current drive for each LED. The transistor is a general purpose NPN device and merely acts as a solid-state switch.

Development idea: A CMOS logic probe would require different levels, nominally 70% and 30% of V_{CC} (worst case 60% and 40%). The 7555 timer (CMOS 555) has thresholds of $\frac{1}{3}$ and $\frac{2}{3}$ of V_{CC}, so it could form an excellent front end for the probe.

Circuit operation

With a low on the probe tip the transistor is off, the inputs to U1b and U1c are low and LED2 is off. The output of U1f goes high, re-inverted by U1d and U1e (in parallel), causing LED1 to conduct. Q2 is a small signal transistor used as a low forward voltage diode. This adjusts the threshold so that marginal low signals are not detected, only stable ones. R1 is then necessary as a collector load.

With a high on the probe tip, Q1 conducts and a high appears on the inputs of U1b and U1c. This high is inverted and LED2 conducts. The output of U1f goes low, re-inverted by U1d and U1e, and LED1 is off.

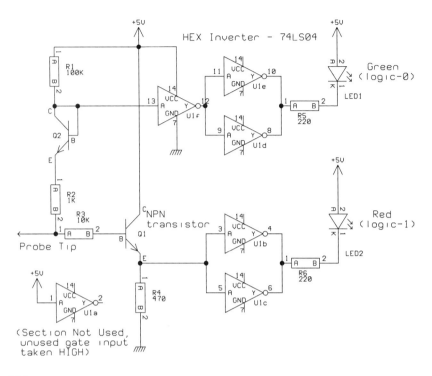

Figure 10.1

With no signal detected at the probe tip, the input to U1f floats high and thus LED1 is off. Both transistors are also off. Because of the way Q1 has been biased, it needs a large positive voltage (a high) to turn it on. Thus the inputs to U1b and U1c are low and LED2 is off.

Construction

Following the stages described in the Practical Exercise in Chapter 9, draw the schematic, capture it and lay out a PCB using the same outline shape.

Troubleshooting

There is not much to go wrong here, but if fault-finding becomes necessary you will need access to a multi-meter or at least something to measure current and voltage. Before going into the circuit in depth, first check the integrity of the supply and see how much current is being drawn. Then check all the components; confirm that the values are correct and that they have been fitted correctly. Check that there are no solder bridges or broken tracks on the circuit board.

Having accounted for about 75% of the possible errors with no further success, you will need to trace the signal back from the output devices to the input. If LEDs do not light up as expected and they are fitted correctly, ground the anode of the suspect device through another resistor. If it does not light up then it is defective, or has been wired backwards. Check each inverter by grounding its input and recording a high (between 2.5 and 5 V) on its output. With a high (5 V) on the base of the transistor its emitter should be at least 4 V.

2. A fluid level indicator

In this example (Figure 10.2), four electrodes are positioned in a container for fluid. Three of these electrodes are connected to the base of transistor drivers, while the fourth is connected to a 5-V supply. The fluid itself acts as a conducting medium such that an indication of its level is displayed on a seven-segment display.

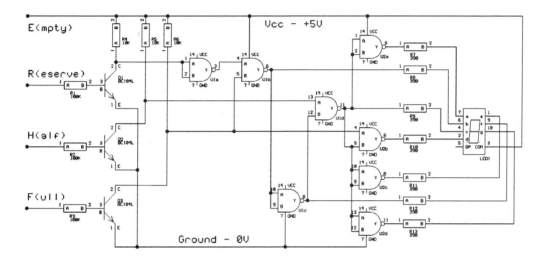

Figure 10.2

Operation

When the fluid level is at or above the full (F) electrode, the emitters of each transistor will be low and the logic will drive four segments to display an F. With the fluid level above half (H), but below full, Q3 collector is high and a 2 is displayed. Above reserve (R) and below half, Q2 and Q3 collectors are high and a 1 is displayed. When the fluid level is below reserve, the collectors of all three transistors are high and an E is displayed.

A logic 0 drives each segment and the logic can be described in a truth-table.

	Q1E	Q2E	Q3E	a	b	c	d	e	f	g
F	0	0	0	0	1	1	1	0	0	0
2	0	0	1	0	0	1	0	0	1	0
1	0	1	1	1	0	0	1	1	1	1
E	1	1	1	0	1	1	0	0	0	0

The table shows that $a = e = g$, $c = a$ and $b = f$.

Note that with the circuit as shown, several gates are duplicated. This is to ensure that no more than one segment is driven by each gate, to avoid overloading the outputs.

Investigate the logic required to display other symbols. In particular, only four of eight possible states have been decoded. If a fault condition existed, one of the illegal states could arise. Detect these, and arrange for a recognizable error display.

3. LED strobe display

This circuit (Figure 10.3) uses a Schmitt gate to generate a clock signal, which with a 74LS193 synchronous 4-bit up/down counter and a 74LS154, 4-to-16 line address decoder, provides a sequence of flashing LEDs. A 74LS00 quad 2NAND provides a flip-flop circuit and there is a small driving circuit for a loudspeaker. As only one LED lights up at a time, just one 470-Ω resister (R3) is needed to provide excitation current.

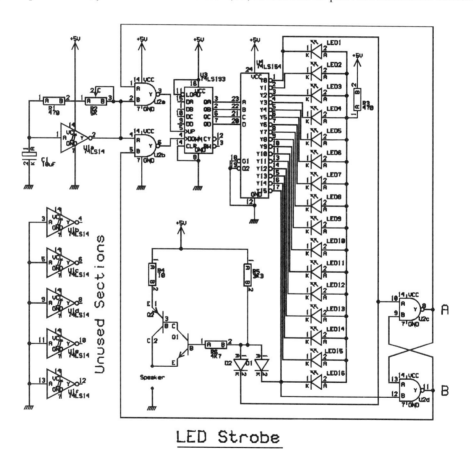

LED Strobe

Figure 10.3

When the output of U2c (point A) is high and the output of U2d (point B) is low, the counter counts up. The decoder provides low states on each of its outputs in turn, starting at Y0. When the low reaches Y15, D1 conducts and the amplifier will deliver an audible pulse to the loudspeaker.

When point A is low and point B is high, the counter counts down. The decoder provides low states on each of its outputs, starting at Y15 and ending at Y0. When the low reaches Y0, D2 conducts and the amplifier will deliver a second pulse to the loudspeaker. Ideally, the speaker should be connected such that no d.c. current flows through it, but this has been ignored in the present case as faithful reproduction of the pulse is not important.

The strobe rate can be altered by adjusting R3.

4. An entrance and exit detector

This is a sensing device that can work with the up/down counter described in Chapter 6. The circuit (Figure 10.4) uses two phototransistors, Q1 and Q2, and a 74LS132 quad Schmitt 2NAND. Q1 is used as the first exit detector and Q2, the first entrance detector and the outputs of U1c and U1d are connected to the up/down control of the chosen counter circuit.

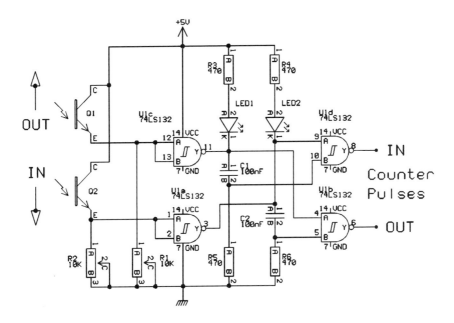

Figure 10.4

Circuit operation

The 1-kΩ presets are adjusted so that when Q1 and Q2 are not blocked from a light source both emitters are high or logic 1. U1a and U1c invert this logic to low, and thus the outputs of U1b and U1d are both high.

When an exit is detected by Q1, its emitter goes low and the input on pin-4 of U1b is high. Q2 is then blocked and with both transistors now blocked the output of U1a changes from low to high, thus providing a positive differential pulse at pin-5 of U1b. The output of U1b delivers a negative pulse to the out counter line.

An entrance is detected by Q2, its emitter goes low and the input on pin-9 of U1d is high. Q1 is then blocked and with both transistors now blocked the output of U1c changes from low to high, thus providing a positive differential pulse at pin-10 of U1d. The output of U1d delivers a negative pulse to the in counter line.

The LEDs give a visual indication when the circuit is enabled.

As can be seen from the above description, the detector will recognize objects that produce phototransistor emitter signals as shown in Figure 10.5(a). Unfortunately, narrow objects could produce valid signals, as in Figure 10.5(b). Also, the illegal signals in Figures 10.5(c) and 10.5(d) should ideally be recognized and ignored. It is quite a challenge to analyse this problem and produce enhanced logic capable of this, especially as it is asynchronous. (*Hint*: It can be solved using just three SR latches, and quite a lot of gating!)

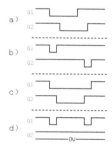

Figure 10.5

Construction

You will need to design both a schematic and PCB component for the phototransistor. The Siemens SFH610-2, supplied by RS as part number 585-214, should serve nicely.

5.   Simple frequency counter

The final circuit is slightly more ambitious and demonstrates the principle of frequency counting. It can also serve as a primitive voltmeter or ammeter. The design of the BCD to a seven segment display interface has been left to you.

Four fixed time-bases are generated by one section of a 74LS14 hex Schmitt inverter and a chosen capacitor. This provides a gating pulse for the decade counter circuit. With an input signal at F_{in} (Figure 10.6) and the time-base signal high, counting begins. On the negative edge of the time-base, the two SN7475 latches are enabled to drive a seven segment display. When the time-base is low, counting stops.

There are two principal disadvantages to this design. The frequency determining components are subject to drift, and the timing capacitors must be in exact decade multiples. A more accurate clock would use a crystal oscillator, with a chain of decade dividers to give other ranges.

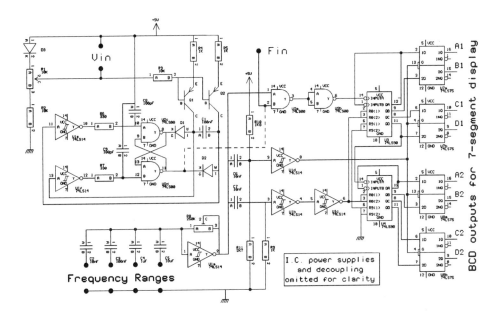

Figure 10.6

The VCO (voltage controlled oscillator), whilst linear, exhibits a d.c. offset due to the transistors V_{be}. It is instructive to plot input voltage against frequency. The effective range is from $V_{CC} - 0.6\,V$ to $V_{CC} - 2.5\,V$. Note the divider formed by R10 and R11 on the input to U1d. This holds the input at about 3 V, so negative pulses through C6 will trigger the inverter.

Index

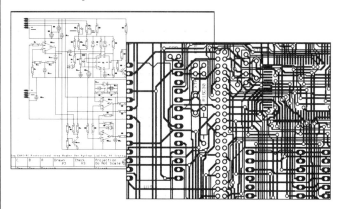
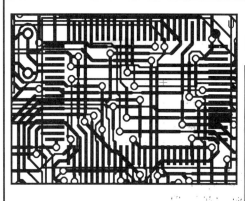